THE SOCIAL HISTORY OF OCCUPATIONAL HEALTH

THE
SOCIAL HISTORY
OF
OCCUPATIONAL
HEALTH

Edited by
PAUL WEINDLING
for the Society for the Social History of Medicine

CROOM HELM
London • Sydney • Dover, New Hampshire

© 1985 Paul Weindling
Croom Helm Ltd, Provident House, Burrell Row,
Beckenham, Kent BR3 1AT

Croom Helm Australia Pty Ltd, Suite 4, 6th Floor,
64-76 Kippax Street, Surry Hills, NSW 2010, Australia

British Library Cataloguing in Publication Data

The Social history of occupational health.
    1. Industrial hygiene – Europe – History
    I. Weindling, Paul II. Society for the Social
    History of Medicine
    613.6'2'094   HD7694

    ISBN 0-7099-3606-0

Croom Helm, 51 Washington Street, Dover,
New Hampshire 03820, USA

Library of Congress Cataloging in Publication Data
Main entry under title:

The Social History of Occupational Health.

    Based on presentations made at a conference held
at Portsmouth Polytechnic from 8–10 July 1983.
    Includes bibliographies and index.
    1. Industrial hygiene—history—congresses.
2. Industrial hygiene—social aspects—congresses.
3. Medicine, industrial—history—congresses.
4. Medicine, industrial—social aspects—congresses.
I. Weindling, Paul. II. Society for the social
history of medicine.
RC967.S58   1986   363.1'1'09   85-21286
ISBN 0-7099-3606-0

Typeset at Oxford University Computing Service
by Helen Jones, with Frances White and Paul Weindling
Printed and bound in Great Britain by
Biddles Ltd, Guildford and King's Lynn

Contents

Part One: Introductory

Part Two: Social Conditions and Risk Factors

Contributors

Mel Bartley, University of Edinburgh, Department of Social Administration, Adam Ferguson Building, George Square, Edinburgh EH8 9LL.

Peter Bartrip, University of Oxford, Centre for Socio-Legal Studies, Wolfson College, Oxford OX2 6UD.

Linda Bryder, University of Oxford, The Queen's College, Oxford OX1 4AW.

Gill Burke, Polytechnic of Central London, School of the Social Sciences and Business Studies, 32-38 Wells Street, London W1P 3FG.

Karl Figlio, Free Association Books, 26 Freegrove Road, London N7 9RQ.

Antonia Ineson, 22 Corinne Road, London N19.

Helen Jones, University of Oxford, Wellcome Unit for the History of Medicine, 45-47 Banbury Road, Oxford OX2 6PE.

Alfons Labisch, Gesamthochschule Kassel, Fachbereich 4/ Sozialwesen, Heinrich-Plett-Strasse 40, 3500 Kassel, Federal Republic of Germany.

Lothar Machtan, Universität Bremen, Fachbereich 11, Bibliothekstrasse, Postfach 330440, 2800 Bremen 33, Federal Republic of Germany.

Dietrich Milles, Universität Bremen, Fachbereich 11, Bibliothekstrasse, Postfach 330440, 2800 Bremen 33, Federal Republic of Germany.

Rainer Müller, Universität Bremen, Fachbereich 11, Bibliothekstrasse, Postfach 330440, 2800 Bremen 33, Federal Republic of Germany.

Deborah Thom, University of Cambridge, Child Care and Development Group, Free School Lane, Cambridge CB2 3RF.

Paul Weindling, University of Oxford, Wellcome Unit for the History of Medicine, 45-47 Banbury Road, Oxford OX2 6PE.

Perry Willson, University of Essex, Department of History, Wivenhoe Park, Colchester, Essex CO4 3SQ.

Illustrations

Preface

Since its foundation in 1970, the Society for the Social History of Medicine has focused attention on a very wide range of social aspects of health and medicine. For instance in 1975 the Society held a conference on 'Industrial Medicine'. Summaries of papers were published in the Society's Bulletin no. 16. The present volume derives from the Society's conference on 'The History of Occupational Medicine'. This was held at Portsmouth Polytechnic from 8-10 July 1983 with the assistance of Peter Wright. Bulletin no. 33 contains summaries of the conference papers. This volume contains a series of studies on the social history of occupational health, as well as providing a review of existing literature and approaches. It is hoped that the volume will stimulate further historical work, as well as showing the relevance of historical issues to present conditions.

It was particularly stimulating to have had the participation of members of the Society of Occupational Medicine and of the Bremen/Kassel project on the history of occupational medicine. The planning of the conference benefited from lively discussions with Rainer Müller while he was based at the Wellcome Unit for the History of Medicine, Oxford, during 1983.

The conference included rural and other non-industrial aspects of occupational health. It was felt that a cohesive volume could be best achieved by concentrating on social aspects of industrial health, and this volume consists of revised papers dealing with the history of industrial health. I am grateful to Helen Jones for providing an additional paper. My introduction has been enriched with advice from Mel Bartley, Helen Jones, Dietrich Milles, Rainer Müller, Margaret Pelling and Charles Webster. The production of the book was itself an adventure in coming to terms with new technology. Much patient advice and guidance was provided by Kathleen Griffen, Susan Hockey and Gayan Mathur of The Oxford University Computing Service, by Alan Scott and Frances White of Corpus Christi College, Oxford, and by Jonathan Barry of the Wellcome Unit for the History of Medicine, Oxford. Helen Jones took a major role in the production of this volume, becoming expert in the use of the Oxford University Monotype Lasercomp typesetter. Ann Cheales provided valuable secretarial assistance. I also wish to thank the Publication Sub-

Committee of SSHM, Virginia Berridge, Ludmilla Jordanova and Adrian Allan (also the Society's Treasurer) for considerable support.

Paul Weindling

FOREWORD

If we are going to achieve a major improvement in health care in this country, we must extend our horizons beyond the conventional limits of the health service. The 1980 Black report on the Inequalities of Health showed how, in spite of the welfare state, class differentials in health still exist. For many workers, accepting a job in industry means that their shortened life expectancy may be further impaired because of the work they do.

While it is true that the more horrific hazards of early industry (such as those described by Engels in his study *The Condition of the Working Class in England*) have been largely eradicated, some of the processes involved in industrial work have remained more or less unchanged. The effects of repetitive manual work, which he characterised as the 'the most deadening, wearying process conceivable', still result in excesssive physical deterioration and mental fatigue. Notwithstanding health and safety legislation controlling the conditions of work, noise and hazardous substances continue to cause chronic disabilities and claim lives. The difference between the industrial revolution and now is that many of today's cases of industrial illness and death are easily preventable. Their continued occurrence amounts to nothing short of criminal neglect and is an indictment both of the low priority given by employers to health and safety at work and the continuing inequality in access to and provision of treatment from the National Health Service.

Nor have adequate compensation provisions emerged as rapidly as they might have; indeed they are now being weakened as a result of changes in the social security system. After suffering an injury at work from the effects of a work related disease, disablement benefit has to be obtained through a damages claim against an employer, or by a claim for Disablement Benefit to the Department of Health and Social Security. There follows what can be a harrowing experience of Medical Appeal Boards, legal wranglings and appeal procedures — with the burden of the initiative on the already injured claimant. If the disease is not already on the all-too-slowly expanding list of Prescribed Industrial Diseases, and if the claimant already has some form of disability, benefit will be unobtainable or curtailed.

The importance of trade union involvement in these processes cannot be too strongly emphasised, and not only in terms of providing guidance and support to individual claimants. If the trade union movement does not get actively involved the industrial health problem is in danger of being made to look smaller than it is — with a minimal public demand for a clean-up.

There needs also to be a far greater recognition of the contribution of labour historians and social scientists. By continually asking 'how effective?', 'how often applied?', 'how did class relations and the distribution of wealth affect specific situations?', history helps us to understand how we have arrived at the current situation of industrial safety standards, factory inspection, compensation provision and safety promotion. History also provides a catalogue of past mistakes, of negligence, and of tolerance of what should not have been accepted as well as examples of successful preventive measures. Of course, the diseases of the past are not always the diseases of the present. The epidemiological pattern is constantly changing. Before the N.H.S. arrived, tuberculosis was a major killer with many occupations bringing increased risk of T.B. Today cancer and heart disease have taken its place. Just as improved social conditions helped turn the tide against T.B., so cancer and heart disease should now also be regarded as preventable social problems with diet, exercise and anti-smoking pressures playing a major role.

The inclusion of international comparisons in this book is particularly welcome. History shows that doctors, public health experts, trade unionists and employers have frequently borrowed from developments in other countries. No international organisation can have contributed more to this collective effort than the International Labour Organisation, which continues to provide vital information about the varying contributions which workers, employers and the state make to levels of care in their different countries. In some countries and at certain points in time, health care was arranged either by workers' friendly societies or through the provision of factory doctors who would treat not only employees but also their families. Such situations contrast sharply with Britain's National Health Service where industrial health services have no part to play. The reasons for this exclusion are important historical questions in themselves, with the validity of the separation of the N.H.S. and the Health and Safety Executive very much a current issue, particularly in view of the present cutbacks in both sectors.

The situation is never static; new technologies can override the accepted norms of health and safety standards. The high accident rate of North Sea Oil exploration and extraction is a case in point. There are often immense human and social costs involved in the handling of toxic substances and of hazardous work situations. Legislation is important, but to be effective needs the co-operation of employers and workers. Even after legislation takes effect, abandoned practices and substances can have a malign effect later on. For this reason it is necessary to monitor the health of occupational groups over long periods. It is also vital that the records of benefit and claims, as well as factory inspection and medical records, are preserved. The Wilson Report on the selection of, and access to, public records has stressed the need to halt short-sighted and ill-considered destruction of medical case histories and public health records. Without these records it is impossible for medical and social scientists to monitor the health of groups of workers and particular diseases over long periods of time.

Social history also shows us that occupational health is not just the concern of industrial workers. Work is, or should be, a common experience. Although far too little research has been undertaken into the impact of unemployment on health, it is increasingly recognised that loss of work is detrimental to health. Society as a whole has a responsibility to ensure that work and its rewards, in terms of adequate levels of pay, job satisfaction, the value of the product, the personal dignity of the worker and the collective experience of groups of workers, bring real benefit to all members of our society — so that the choice of work is an enriching rather than degrading experience that is available to everyone. Industrial health is not just about 'preventive medicine' based on the physiology of the worker. It encompasses a whole range of economic, social and specifically personal factors. It is relevant to unwaged work, including housework and child rearing as well as the multiple occupations that are part of rural economies. It means prevention of the production of commodities — from nuclear weapons to cigarettes — that damage the health of producers and consumers alike and the development of socially productive alternatives. These changes require a shift not just to medical but also socio-economic strategies of prevention that would improve the overall quality of our people's health and make the conclusions of the Black Report a thing of the past.

Michael Meacher M.P., Chief Opposition Spokesman on Health and Social Security.

Part One: Introductory

I LINKING SELF HELP AND MEDICAL SCIENCE: THE SOCIAL HISTORY OF OCCUPATIONAL HEALTH

Paul Weindling

Occupational diseases and disabilities provide a sensitive index of social conditions in industrial societies. Yet at times occupational health and medicine can be ranked as among the most depressed areas in industry and medicine. Only a fraction of the historical literature on industry, the labour movement and medicine is concerned with occupational health. This book aims to remedy neglect of occupational health, and establish it as part of the social history of industrialisation.

The case-studies presented here have been restricted to industrial health as there is a dearth of historical work on rural health. In rural and pre-industrial economies the relations between the domestic economy of the family and occupations, often multiple, are complex and less subject to formal definitions of the scope, hours and duties of work. Also excluded are those groups like soldiers or school children who have benefited from especially comprehensive health services, and domestic work for which health provisions have been conspicuously absent.

By taking health and labour as the focus for a wide range of social factors, the contributors do more than provide an account of scientific progress in occupational medicine. In the wake of industrialisation it was hoped that promotion of scientific approaches to occupational health would establish safe standards, and provide humane, legally just and medically beneficial solutions to problems of industrial diseases and disabilities. Yet industrial hazards pose complex problems, which can be seen from many different perspectives. Mediating between the interests of worker and employer are state and local authority factory, medical and environmental health inspectors, party political and government agencies, judicial and statutory procedures, and a range of professional experts like general practitioners, industrial medical officers and nurses, industrial hygienists and toxicologists, psychologists and safety officers. A worker's sense of aches and pains might not accord with what can be recognised as a compensatable disease by the complex legal and medical machinery.

It has been suggested that diseases and disabilities have a 'dual nature', one emotional, personal and political, the other scientifically and legally rational. However desirable a value-neutral and technocratic approach might be, historical reconstructions of scientific work show widely differing perceptions of diseases and health hazards. Laboratory investigations have often been difficult to apply in practice. Contributions from the Federal Republic of Germany introduce the concept of *Dethematisierung*, or suppression, to express the loss of social perspectives in scientific occupational medicine. It is necessary to restore awareness and 're-thematise' how occupational health has been affected by a very wide range of social factors.[1]

I Towards a Social History

Textbooks of occupational medicine have often included lengthy historical sections as a way of introducing medical students to the social problems. Nearly one third of Donald Hunter's *The Diseases of Occupations*, which was first published in 1955, consists of an historical survey. Hunter's approach was distinctive for its emphasis on clinical case-histories: it was less geared to legislation than earlier writings, and less epidemiological and toxicological than later works. Hunter included rare and obsolete diseases as historical case studies in order to warn clinicians of the limitations of medical science, particularly in the face of social problems. Yet Hunter's use of history was limited to progress in legislation and in medical research, and neglected problems of implementing safety measures. The positivistic faith in progress of the medical sciences lacked awareness of how science itself can represent a type of vested social interest which must be subjected to critical scrutiny.[2]

Since the 1930s there have been demands for a social history of medicine, taking account of occupational conditions. Alfons Labisch's chapter shows how diverse approaches to the history of occupational health have co-existed in German-speaking countries. At one extreme there has continued to be a vein of positivistic and narrowly focused professional eulogy. But there has also been the attempt to reorient medicine to social problems by emphasising the social history of medicine from the patient's point of view. This approach was developed by Henry Sigerist at Leipzig and then at Johns Hopkins University, Baltimore during the 1930s. Sigerist campaigned for and studied industrial health insurance schemes.[3] The new attitude in

Germany impressed a visiting American researcher, George Rosen, who called for the history of occupational medicine to become a major branch of history. In 1937 he predicted that the 'New History' of social conditions and culture, combined with the awareness of the importance of social medicine, would result in development of the history of occupational medicine.

During the 1920s there were major initiatives to restore social and political dimensions to occupational medicine by the first medical inspector, Thomas Legge, and the pioneer of social medicine, Ludwig Teleky.[4] Rosen's programme was designed to underpin such a reorientation of attitudes, along with the campaigns by Sigerist for socialised medicine. Rosen condemned the narrow study of scientific advances in physiology and toxicology. Instead, he hoped that study of occupational diseases in their social context would open new ideological perspectives, contribute to the solution of problems in industrial medicine, and broaden understanding of social conditions. Despite Rosen's achievements with his *History of Miners' Diseases*, his history of public health, and in his classic essays *From Medical Police to Social Medicine,* his call fell largely on deaf ears. Although the Second World War saw the increased importance of social medicine, the Cold War and scientific priorities blocked development of social approaches in medical history.[5]

In Britain the wartime improvements in occupational health and social medicine failed to culminate in a state occupational health service. A select few in occupational medicine continued to produce historical studies of lasting value, and oriented to current problems of occupational health. The Glasgow expert in public health, Thomas Ferguson, made substantial contributions to the impact of industrial work on sickness in Scotland. Andrew Meiklejohn, a specialist in silicosis, had a keen sense of how history was relevant to current problems in occupational medicine. He saw how since the 1930s the shift in research on silicosis, from physiological approaches to the recognition that it was caused by dust rather than muscular exertion, marked a return to views of the 1860s. When a new initiative on occupational health was taken by the Ministry of Labour in 1955, he argued for a revival of the 'revolutionary' ideas of Thomas Percival and the Manchester Board of Health of the 1790s. He recognised the advantages of earlier more environmental approaches to medicine.[6]

Precursors of the social history of medicine are to be found among those pioneering reformers and social scientists who examined health as part of more general assessments of standards of living and of demographic change. Government commissions and social commentators like Henry Mayhew, Charles Booth, Beatrice and Sidney Webb, G.D.H. Cole and Seebohm Rowntree used medical and nutritional evidence with regard to industrial welfare. B.L. Hutchins' *Women in Modern Industry* of 1916 included much historical material, arising from an earlier study on labour legislation and from first-hand investigation of the strain produced by long hours, low pay and insanitary environments. Historians of the industrial revolution placed a new emphasis on the workers' point of view. This is exemplified by the works of J.L. and Barbara Hammond, who in 1911 portrayed the village labourer and in 1917 the town labourer. Whereas the Hammonds and Alice Clark regarded the industrial revolution as producing a worsening of labouring conditions, Ivy Pinchbeck argued in 1930 that women workers benefited from greater freedoms and prosperity of industrial society. Dorothy George vividly depicted the working lives of eighteenth-century Londoners prior to the Health and Morals of Apprentices Act of 1802, linking eighteenth-century concern with the parish poor to nineteenth-century factory regulations. Mabel Buer's 1926 study of the health of the population from 1760 to 1815 looked at the relation between working conditions and health in order to ascertain the human cost of economic change. Yet these authors gave more attention to housing, nutrition and income levels rather than to the direct ill-effects of the labour process. Modern commentators on the welfare state like Richard Titmuss continued to be more interested in the relations between the factory and society, rather than in conditions at the workplace. Their omission has reflected the exclusion of occupational health from the National Health Service. The social science tradition from the early social surveys to modern commentaries on the welfare state was the source of the social history of medicine.[7]

In Germany there has been a parallel tradition of social scientists as commentators on working conditions. A stimulus to study of occupational conditions was given by the associations for social policy and for social reform active during the 1880s and 90s. The pioneers of sociology, Alfred Weber and Max Weber, were interested in workers' living conditions and industrial psychology. Labisch shows how diverse levels of inquiry into occupational health have been sustained by social scientists. As in Britain health is considered as a major factor in social conditions in association with other variables like income, diet, housing and education. Comparison between Germany and Britain has been largely on the level of legislative provisions. The perspectives developed by Milles suggest that comparison should also be made with regard to workers' attitudes to diseases, to the implementation of preventive measures and to the working of compensation claims. Any attempt at comparison has to take the underlying social conditions into account, as well as differing registration procedures. Small-scale case studies have considerable advantages of authenticity over large aggregates of national statistics.[8]

Deriving from the work of commentators on social conditions are a number of further sources for the social history of medicine. Demography raises questions of the effects of industrialisation on the birth rate, on migration and on mortality. Medical concerns with the incidence and treatment of disease are central to demography. Studies of the intellectual repercussions of industrialisation provide a context of religion and secularisation for the growth of new attitudes to labour and health, and for the rise of science and technology. The links between the rise of capitalism and a new positive attitude to the dignity of manual work as state of grace and a means of human salvation, as suggested by R.H. Tawney and Max Weber, pose questions of the predisposing outlook enabling occupational health to become a matter of social concern. Doctors' writings on working conditions stemmed from the 'history of trades' genre originating in the scientific revolution of the sixteenth and seventeenth centuries, and reflected broader philanthropic and humanitarian social attitudes. Industrial welfare schemes and model communities drew on ideas developed in religious and technological utopias since the Renaissance.[9]

The social history of medicine can make a positive contribution to studies of intellectual and social change, by taking an innovative lead on neglected social issues. One area meriting increased attention is that of chronic diseases and disabilities. These can be as significant as major mortality crises like epidemics, or major accidents in mines and factories when whole communities are devastated, as with the estimated 2500 killed by leaking methyl isocyanate at Bhopal in 1984. Certain chronic diseases and disabilities have a major debilitating effect on everyday life, although they cause only a low proportion of deaths, or are only indirect causes of other fatal or disabling conditions. Examples of chronic conditions with such a low mortality are venereal disease, alcoholism, intestinal parasites, and tooth decay. From the 1890s there grew a substantial literature on their effects on industrial efficiency. Industrial work can cause deafness, skin conditions, fractures (with possible long term effects), back pain, hernias due to lifting heavy loads, arthritis, rheumatism, and varieties of poisoning due to toxic substances. Many chronic conditions can be made worse by the stresses and strains of work. In 1945 Thomas Ferguson criticised the scientific bias towards toxicology and pathology, so as to emphasise the extent of chronic diseases. He described industrial health as 'really a tale of men and women bronchitic, rheumatic, cardiatic, spent ere their time'.[10]

There has been only limited use of accidents and diseases as an indicator of social conditions. Much can be learnt from medical institutions like hospitals and facilities like dispensaries and welfare clinics for outpatients many of whom suffered from accidents or chronic conditions. There are few long-term studies of sickness as related to industrial work. Nor has there been any sustained attempt to locate and analyse the records of the early works doctors. Routine medical examinations are a neglected historical source, and the discussions of 'malingering' early this century reveal a critical attitude of doctors to the reported symptoms. Sickness absence statistics have been analysed by Ferguson for the Deanston Mill from 1829 when a factory doctor was appointed until 1832.[11]

Attitudes to the deformities and diseases of workers have betrayed on occasions racial bias. During the nineteenth century it was feared that industrial living conditions would produce an inferior race of urban degenerates. Analogies were made between paupers and savages, providing an inspiration for Mayhew's ethnology of working types. Eugenicists were among the most perceptive commentators on chronic diseases, stigmatising TB, VD and alcoholism as 'racial poisons'. Eugenicists were especially concerned that women's work could cause physical degeneration and inherited malformations, and at least would divert women from their natural role as mothers. Ineson and Thom provide a striking analysis of women munitions workers' chronic ill-health with a low mortality but a high degree of suffering. While there was mass slaughter among the men at the Front, women industrial workers were enduring exposure to hazardous and toxic substances. The relations between women's work, health, pregnancy, child care and the division of labour in the family raise fundamental issues in industrial society.[12]

Work also relates with standards of sanitation and cleanliness. Burke points to the risks of ankylostomiasis or hookworm infection in mines due to insanitary conditions. Hookworm was called 'the germ of laziness' when its debilitating effect on workers in the Southern USA was shown in the 1890s.[13] Other categories of diseases have been caused by malnutrition, exacerbated by heavy manual labour. Scurvy was worsened by the harsh working conditions on ships, and could cause the death of a high proportion of ships' crews. Ships' officers were regarded as less prone to acute scurvy owing to their lighter duties. Leg ulcers were a major complaint at the dispensaries of the early industrial revolution, owing to poor nutrition and the strains of industrial labour. Some afflictions have disappeared from modern medicine, like 'chlorosis' or 'the green sickness', which were once common diagnoses for exhausted and undernourished women factory workers. Other disabilities have all too often been medically underestimated until recent years, one example being deafness due to excessively loud machinery: the debate still continues over what constitutes a safe level of noise control. Asbestosis and radioactivity have been classic examples of underestimated industrial hazards, until recently largely overlooked even though risks were medically first demonstrated early this century.[14]

The emphasis on chronic diseases in the social history of medicine is also justified in that chronic diseases like tuberculosis caused many deaths. Cancer and heart disease have in modern industrial societies replaced TB as major causes of mortality. Bryder's contribution deals with TB as an occupational disease. The case study of slate workers in North Wales has broader relevance in that it shows how the decline in TB may be regarded as due more to changes in social and working conditions than to improvements in medicine. Bartley's chapter raises the issue of the extent to which disadvantaged social classes have been at greater risk from heart disease since the 1920s. Debate still rages today over the proportion of cancers which are caused by industrial products and processes, as opposed to personal life style with the effects of smoking, diet and sexual behaviour.[15]

There is an important place for historical discussion of the impact of socio-economic conditions on disease and death rates with morbidity and mortality taken as indices of this impact. Many aspects of major changes in mortality patterns remain as yet inadequately explained, for example the reasons for the fall in infant mortality from high levels in the 1900s, and for the decline of tuberculosis. Added to these problems should be the role of occupational factors as causes of death and disease in industrial societies, either in a direct sense as causes of specific diseases or disabilities, or indirectly, owing to low wage levels, excessive working hours, poor housing and environmental pollution.

II Responses to Risks

Diseases have often been identified by specific industries, manufacturing processes, or by symptoms. It was only during the nineteenth century that advances in pathology and bacteriology shifted the emphasis away from environmental and symptomatic descriptions to classifications based on infecting agents. There have consequently been a great many local descriptions of diseases which it is difficult to relate to conventional medical classifications. This has caused problems in diagnosis and in compensation claims, and widened the gap between diagnosis and therapy. Many contrasts can be shown between workers' attitudes to diseases, and medical diagnoses. Retrospective diagnosis presents the medical historian with many problems.

Diseases were given names by the workers themselves, who took the risks into account. During the nineteenth century the diseases of grinders were recognised as a major hazard, and were described in evocative terms as 'spit', 'lung', 'rot' or 'asthma'. Attempts to introduce safety appliances could not alleviate the economic pressures at work, and might slow down piecerate workers. The Sheffield grinders of the 1860s recognised the risks, but felt that the economic system was at fault:

> He shortens his life, and he hastens his death
> Tally hi-o, the grinder!
> Will drink steel dust in every breath...
> Won't use a fan as he turns his wheel
> Won't wash his hands ere he eats his meal
> But dies as he lives as hard as steel...
> Where rests the heavier weight of shame?
> On the famine-price contractor's head
> Or the workman's under-taught and fed
> Who grinds his own bones and his child's for bread?[16]

Far too little is known about workers' perceptions of industrial risks. The element of danger could mean that a hazardous trade could command higher rates of pay. An important feature of Burke's chapter is that it brings out the miner's response to the certainty of premature death, which was weighed against compensation. Milles comments on the way in which the worker took diseases into account, when balancing the risks of unemployment and starvation against the physical costs of an occupation which was likely to prove fatal. Workers were often reluctant to admit that they were sick until forced to, as serious or frequent illness could mean dismissal.

Medical supervision and regulations have been subject to much variation owing to social circumstances. Ineson and Thom describe the reactions of women munitions workers to skin discolouring and other illnesses resulting from exposure to toxic substances. They point out that medical standards were deliberately lowered for munitions workers. An analogous case study has been made of North Sea oil exploration by Carson. This suggests that safety could have been much improved by enforcing existing regulations, but the regulation of safety has been downgraded owing to the nation's economic priorities.[17]

There has been a long tradition of workers' self-help with workers able to treat more common injuries. This has found more formal expression in voluntary first aid and rescue services. Attempts to alert workers as to the risks of their occupations used education in promoting self help. For example, in 1825 a tract warned painters and varnishers of the risks of their trade.[18] At times promoting self help has been used to shift the burdens of blame and responsibility solely on the worker. The most important example of self help was that of friendly societies' sickness benefits during the nineteenth century, and many unions undertook friendly society activities.[19] Trade unions have often made pay a priority, and neglected health issues. However, the Trades Union Congress and unions have developed considerable expertise in the pursuit of compensation claims. Compensation and liability place the burden of establishing a claim on the shoulders of the injured party, as compensation is not paid on the basis of need. Bartrip reviews compensation legislation, and Figlio unpacks the conceptual complexities of legal claims procedure with diseases reduced to the status of accidents, and accidents precipitating wrangles over liability.[20]

Nineteenth-century public health experts like Chadwick and Simon recognised that it was most appropriate for medicine to intervene where it interfered least with vested interests. Such considerations have resulted in compensation being preferred to preventive public health measures. Compensation has required objective medical and scientific standards. It has not always been possible to preserve the image of impartiality in the eyes of the workforce, who could identify the doctor with management. Certification of illness for compensation tribunals could be a cause of grievance, when occupational causes of disease were denied even though a disease might be recognised in legislation as compensable. If a direct cause resulting from occupational conditions could be medically certified, this meant compensation might be paid, but if the causes were associated with poor housing or nutrition — the result of low pay, and so indirectly caused by work — then there could be no compensation. Complex arguments arose over health prior to starting work, over normal and abnormal strains and hazards, and over medical diagnoses and theories.[21]

Occupational hygiene has developed as a distinct field, cultivated by scientists, doctors and trade unions. The hygienist is required to set and maintain objective standards. The International Labour Office owes its origins to international solidarity among trade unions, endeavouring to establish common standards early this century. Unions have become increasingly concerned with hazards at work in order to prevent accidents and minimise the risks of long-term exposure to dangerous substances. The right of the worker to be informed of the risks of dangerous substances and processes is being recognised. Unions have resumed a late nineteenth-century tradition by undertaking their own medical and scientific investigations. *Hazards Bulletin* (1976-1984) of the British Society for Social Responsibility in Science provided an alternative scientific forum, and is itself a useful historical source for the past decade. This newsletter (replaced by *Hazards* and primarily designed for safety 'reps') has hammered home the point that there is a major gap between scientific proof of a hazard and its elimination, and that scientists can themselves disagree over 'safe' and 'hazardous' standards. The recent publicity given to the campaigns over the safe handling and removal of asbestos, and over the risks of radioactive substances provides a measure of the politicisation of occupational health.

There has been surprisingly little knowledge of the amount and types of sickness in industry. Different industrial sectors have different requirements for notification, and the system has a strong voluntary element. Notifiable and compensable industrial diseases constitute only the tip of a submerged mass of illness. The full spectrum of accidents has also been difficult to assess, because statistics of notified accidents, of compensation claims, and of Department of Health and Social Security payments have been separately kept.

Problems of occupational health have been slow to find a substantial response among the medical profession, even though medical research and the tending of the sick are hazardous and stressful occupations. Some medical practitioners' experiences in treating miners and other workers shaped the attitudes of pioneers of occupational medicine.

The priorities of sea-power meant that during the eighteenth century a number of classic studies of the health of seamen were produced, such as James Lind's work on scurvy. Industrialisation generated a substantial medical literature on the dangers of operating machinery.[22] Medical commentators on industrial ill-health were affected by broader currents of social reform. Shortly after philanthropic investigations into the condition of climbing boys, in 1775 Percivall Pott demonstrated how chimney sweep's boys suffered from cancer of the scrotum.[23] Industrial and political upheavals prompted medical practitioners to consider the health implications of working machinery. Workers claimed that large, hot mills generated 'factory fever'. When an epidemic broke out at Radcliffe in 1784, physicians, including John Ferriar and Thomas Percival, blamed the working and living conditions, and apprentices' hours of work were restricted. Ferriar and Percival became active radicals, and supported the Manchester Board of Health to supervise housing and working conditions.[24] Charles Turner Thackrah in Leeds published in 1831 on trades as affecting 'health and longevity'. James Phillips Kay (Shuttleworth) depicted the physical condition of cotton spinners in 1832, and Peter Gaskell wrote in 1836 on the effects of working machinery. Although their work came to the attention of royal commissions and of social commentators like Chadwick and Engels, the Factory Acts allowed for only minimal medical inspection. John Simon, after Chadwick's resignation from the General Board of Health in 1855, directed the attention of medical officers to the ill-effects of occupations. In the early 1860s Edward Headlam Greenhow investigated connections between occupation and respiratory diseases, Augustus Guy reported on arsenic, John Bristowe reported on match factories, William Ord on dress-making and Edward Smith on tailoring and printing.[25]

Occupational health during the nineteenth century was really a part of public health. In the twentieth century the up-grading of occupational medicine into a postgraduate specialism has been achieved by emphasising scientific priorities. Concern over the health of women munitions workers, as described by Ineson and Thom, provides the background to the establishing of the Industrial Fatigue Research Board. The Medical Research Council took over responsibility for this Committee, which became the Industrial Health Research Board in 1928. The physiological approach provided opportunities for leading scientists like J.S. and J.B.S. Haldane, and Charles Sherrington, but it had severe limitations. Later MRC initiatives were more successful as with silicosis research, and with the establishing of Donald Hunter's department at the London Hospital in 1943. The London School of Hygiene has a long-established interest in occupational medicine, and the first (but short-lived) department for occupational medicine was founded in Birmingham in 1934. The Nuffield Foundation supported departments in Manchester, Glasgow and Newcastle. The TUC established the TUC Centenary Institute for Occupational Health at the London School of Hygiene.[26] Although it was expected that the university departments of social medicine should take major initiatives in industrial medicine, social medicine has become absorbed in statistical epidemiology and has lost contact with many dimensions of industry.

It has long been recognised that doctors ought to have first-hand knowledge of the workplace and of technical processes, but teaching in occupational health is not a standard part of medical education. There is a preference for treating disease in the context of the family rather than in the context of work. The medical ethic of self-employment resists the linking of practice with industries. The level of interest in preventive methods indicates socio-economic priorities. After cancer of the scrotum in chimney sweeps' boys declined owing to legislation on child work, the disease became common among cotton workers owing to the handling of bituminous substances. Doyal and Epstein point out that this cancer was only controlled during the 1950s when the cotton industry had begun to decline. Mechanisation of mining brought increased incidence of lung diseases owing to the increase of dust. Doctors could also be mistaken in their diagnoses owing to contemporary medical theories. Mass and Levenstein have shown how textile workers contracted byssinosis from having to thread shuttles by sucking the thread, but that at the time this was diagnosed as tuberculosis. Major problems remain regarding the extent to which diseases are a result of occupational factors.[27]

Müller demonstrates the urgent need for long term studies of the health records of occupational groups. Epidemiological monitoring of case records is of major medical importance, as well as yielding further materials for the social historian. In this context it is vital to ensure the preservation of clinical and pensions records, so that working conditions and family background can be correlated with diseases and therapy as well as with details of premature invalidity and chronic diseases.[28]

The substances handled in particular occupations formed the basis of the Registrar General's statistics, as compiled by the Victorian statistician, William Farr. The National Registrar's statistics (today produced by the Office of Population Censuses and Surveys) are an outstanding historical source, permitting correlation of mortality with occupation. They can be supplemented by the statistics of medical officers of health (now community physicians), and by the investigations of research bodies like the Medical Research Council.

Caution is needed in the use of statistical evidence. Bartley points to many problems in the use of such statistics. Clinicians who sign death certificates may not be aware of occupational factors. The subsequent coding of the social class of specific occupations can vary. Government statistics may not show such untidy phenomena as seasonal work, multiple occupations, unpaid or undeclared work, or distinguish between longer and shorter terms of employment. Long-term sickness absence rates can conceal unemployment. Medical officers' statistics were prone to underrate the incidence of sickness, and were more indicative of diversities in medical standards than of differing occupational and socio-economic conditions. Vulnerable social groups may be submerged by larger aggregates of regional and national statistics. This again illustrates the discrepancy between actual experience, and medical and official perceptions. [29]

III Inspection and Community Health

The transition in public health conceptualised by Rosen as the shift 'from medical police to social medicine' applies particularly well to occupational medicine. The idea that a medical police was necessary to enforce effective public health left its mark on the factory inspectorate as a type of industrial police force with powers to prosecute under criminal — rather than civil — law. Machtan and Milles examine the political interests in implementing inspection in Germany, and the barriers to allowing inspectors comprehensive powers of investigation. One of the major criticisms of industrial inspectorates has been lack of personnel, so resulting in a gap between legislation and enforcement. In Britain, contrary to the assumption that there was a low rate of prosecution, there was initially a relatively high rate during the 1840s and 50s. Carson considers that the inspectorate abandoned its policing role during the nineteenth century, although Peacock has argued that the Factory Acts resulted in successful prosecutions from 1833 to 1855. Bartrip examines the changing scope of the inspectorate's responsibilities from the 1880s in the light of changes in workmen's compensation. Jones shows how by the twentieth century the inspectorate felt uncomfortable with their powers of criminal prosecution. When social medicine in the 1920s and 30s turned towards strategies of health education rather than to direct intervention in social conditions, the inspectorate also adopted an educative role. The Home Office Industrial Museum was established as 'A permanent exhibition of methods, arrangements and appliances for promoting safety, health and welfare of industrial workers'.[30]

Medicine has made only slow progress in the official structure of industrial inspection. The Factory Act of 1833 allowed for a surgeon to certify that a child was of the strength and appearance of a nine year old. In 1844 the certifying surgeon was required to examine fitness and accidents, but was appointed by the inspectorate as a result of fears that surgeons might not be objective. Robert Baker was the first doctor to be promoted as inspector in 1858. During the 1870s there was increasing concern with industrial diseases, and with improving the design and ventilation of workshops. In 1879 the Chief Inspector of Factories recognised 'Occupations Injurious to Health' in his Report. Three years after women inspectors were appointed in 1893, a physician, Arthur Whitelegge, was appointed Chief Inspector of Factories, and in 1898 Legge became the first Medical Inspector.[31]

The role of medical practitioners in schemes instituted by employers has a long but sporadic history. Medical attendance was provided for miners at Tintern, sometime between 1575 and 1600. Early records exist of local authorities paying for treatment of accidents at work. In the seventeenth century the East India Company appointed ship's surgeons (but 'factory' surgeons in India were responsible for the health of a whole colonial settlement). The Chatham Dockyard operated a form of medical insurance from the 1590s, and from 1625 a barber-surgeon was appointed full time to the yard. In the eighteenth century medical practitioners were retained jointly by the firm and the workers in a model scheme at the Crowley Iron Undertakings in Sussex. The Quarry Bank Mill of the Gregs at Styal, Matthew Boulton's Soho Manufactory, and in the 1820s the Deanston Mills, employed medical practitioners. The Post Office Medical Service began in 1855 when a full time medical officer was appointed. Railway, gas, canal and Quaker confectionery companies established medical facilities. Manufacturers like Boots, Brunner Mond (later ICI), and Pilkingtons followed these leads, although others like the Morris Motor Co. only instituted measures when there was American pressure. In contrast to later developments, such schemes could provide care not only for workers but also their dependants.[32]

During the twentieth century factories have assumed a greater responsibility in maintaining standards of health and safety. Management-funded employee welfare schemes spread during the 1920s, and were encouraged by the Industrial Welfare Society. Arising from this Society's Advisory Medical Board, the Association of Industrial Medical Officers was founded in 1935. The Second World War saw the acceptance of an industrial medical officer as necessary for larger factories. Yet there were inherent conflicts of interest between industrial medical officers, and general practitioners, who were performing statutory duties of the certifying surgeon. On 25 June 1940 Bevin's policy was accepted that the factory inspectorate should require that larger factories appoint a medical officer, nurses and first-aid staff to supervise health and welfare. Smaller places of work were left without provision, although the late 1940s saw pioneering experiments as at Slough, Central Middlesex, Harlow, and, in the USA, at New Haven for small firms to share medical facilities.[33]

A barrier to any comprehensive occupational health service on a national basis has been the division of health responsibilities between ministries. Despite the energy of the Ministry of Labour under Bevin in expanding occupational health services during the Second World War, integration with the National Health Service was not achieved. School health services were not integrated into the National Health Service for similar reasons. During the 1950s TUC pressure revived interest in occupational health in the Ministry of Labour, but reforms then encountered the government's need to balance welfare expenditure against proposed tax cuts. A further problem has been that there were several agencies for inspection, resulting in uneven implementation of standards and different procedures. In 1974 the establishing of the Health and Safety Executive marked a stage in securing greater co-ordination.

The gap between awareness of hazards and effective action to eliminate the danger is a recurrent feature of the history of occupational health. Lead poisoning is a prime example of a hazard which was long recognised by workers and the government, but over which little was done until the twentieth century. The government from the 1860s aimed to reduce the impact on third parties, a concern which arose from the defence of private property. In the 1920s the National Association of Master Painters and Decorators exerted pressure on the Home Office for increased controls, when crucial evidence was ignored. Rowe's history of lead manufacturing concludes that the lead manufacturers managed to ride out the major storm of controversy without reducing their production.[34]

There have been many initiatives by workers to provide medical care. While Robert Owen's New Lanark Mills were comparable to paternalist experiments of other industrialists who also organised education, sanitary improvements, and sick funds, Owen was remarkable in inspiring co-operative schemes by workers. During the nineteenth century friendly societies' sickness benefits could be supplemented by payments for medical attendance by surgeons who would be paid a capitation fee. A notable example of an extensive medical service controlled by workers was that of the South Wales miners. During the 1930s the TUC allied with the British Medical Association in pressing for better medical facilities for the treatment of industrial injuries. The provision of special accident and orthopaedic clinics gained important union support. Among independent initiatives, the Manor House Hospital, originally founded in 1916 for the treatment of disabled soldiers and sailors, became a centre for industrial orthopaedics, and had a mass membership of penny-a-week subscribers. The need to rehabilitate war wounded, the industrially injured, and those recovering from tuberculosis resulted in innovative schemes in the 1920s and 30s.[35]

From the above survey it can be seen that there have been varied solutions to occupational health in industry. There have been worker or employer controlled schemes, as well as a range of state methods from policing by inspectors to the demand for a comprehensive state occupational health service with preventive and therapeutic responsibilities. Willson's study of the instituting of health and safety councils adds the important dimension of worker participation to the promotion of health and safety — and productivity. The case-study of how fascist ideology and US managerial techniques were linked provides insight into twentieth-century strategies of promoting industrial harmony. Fascist occupational health measures combined certain improved benefits with savage restrictions on workers' rights and the increased hazards of war economies. Nazi Germany legislated for several more occupational diseases to be recognised for the purposes of compensation in the occupational health code of 1936. But in other ways there was a tightening of supervisory controls. The full-time industrial medical officer was a favoured measure. A new cadre of factory doctors were recruited from the elite of Party members in the medical profession. They were partly intended to discourage workers from registering as sick in order to maintain productivity in a war economy.[36] Extermination by forced labour marked the culmination of Nazi policies.

Industrial psychology has marked a further extension of scientific models into the workplace. With regard to positive testing of aptitude and assessing the negative effects of stress, psychology has a major place among the concerns of occupational health and among the techniques of management. More efficiency and productivity were also sought by the new rational scientific systems of labour organisation of Taylor, Ford and Bedaux. Different formulae for de-skilling, developing the division of labour in the context of mass production, and schemes to promote a sense of responsibility have vied with each other.[37]

Yet the tentacles of occupational health stretch even further than the psychology of work. The experience of work has shaped perceptions in everyday activities beyond the workplace. E.P. Thompson has shown how during the industrial revolution industrial work discipline imposed a new sense of time. Certain employers insisted on particular types of religious observance, influenced workers' political affiliations, and curbed the power of trade unions. Tied housing and model industrial communities represent the culmination of the factory's control of workers' lives. Patrick Joyce has described how in Lancashire textile manufacturing towns, factories generated a formal and informal social life, which united work and community life. Washrooms, canteens, and recreational and educational facilities like mechanics' institutes were provided. These served to inculcate thrift, respectability, sobriety and deference, as well as to shift the burden of responsibility for ill-health on workers' lack of cleanliness. Major features of the rhythms of modern life appeared with the tea break (in part to prevent alcoholism), factory canteens (mainly since the Canteens Order of 1940), the weekend, and holidays. The provision of facilities like pithead baths for miners and of other sanitary facilities at work prevented disease as well as changing the daily routine.[38]

These examples show the power of the labour process to structure the everyday reality outside the workplace. Disease and disabilities are not only the most acute expressions of social inequality; they also have a general effect on living conditions. The understanding of occupational health is inextricably related to broader socio-economic conditions in industry. This poses a challenge to historians, who have tended to use official statistics or clinical evidence without taking account that these can underrepresent sickness. History has an important role in reconstructing the changing patterns of diseases and disabilities in society. If the studies in this volume have gone some way to establishing the interdependence of health and work, then it is hoped that this relationship can be shaped in a positive rather than negative way.

Notes

1   G. Rosen, *The History of Miners' Diseases. A Medical and Social Interpretation* (New York, 1943), p. 457; R. Müller and D. Milles (eds.), *Beiträge zur Geschichte der Arbeitererkrankungen und der Arbeitsmedizin in Deutschland* (Düsseldorf, 1984); A. Meiklejohn, 'Doctor and Workman', *British Journal of Industrial Medicine,* vol. 7 (1950), pp. 105-116; K. Figlio, 'How Does Illness Mediate Social Relations? Workmen's Compensation and Medico-Legal Practices, 1890-1940', in P. Wright and A. Treacher (eds.), *The Problem of Medical Knowledge* (Edinburgh, 1982), pp. 174-224; K. Howells, 'Victimisation, Accidents and Disease', in David B. Smith (ed.), *A People and a Proletariat. Essays in the History of Wales* (Pluto Press for Llafur, London, 1980), pp. 181-198; D.M. Fox and J.F. Stone, 'Black Lung: Miners' Militancy and Medical Uncertainty, 1918-1972', *Bulletin of the History of Medicine,* vol. 54 (1980), pp. 43-63; N. Jewson, 'The Disappearance of the Sick Man from Medical Cosmology, 1770-1870', *Sociology,* vol. 10 (1976), pp. 225-244; V. Navarro, 'The Labour Process and Health. A Historical Materialist Interpretation', *International Journal of Health Services,* vol. 12 (1982), pp. 5-29; R.H. Elling, 'Industrialisation and Occupational Health in Underdeveloped Countries', in V. Navarro (ed.), *Imperialism, Health and Medicine* (Pluto Press, London, 1982), pp. 207-234.

2   D. Hunter, *The Diseases of Occupations,* 6th edn (London, 1978); Hunter, *Health in Industry* (Harmondsworth, 1959). For a sample of historical studies by occupational physicians see T. Oliver (ed.), *Dangerous Trades. The Historical, Social, and Legal Aspects of Industrial Occupations as Affecting Health* (London, 1902); Oliver, *Diseases of Occupation* (London, 1908), pp. ix-xix; S.A. Henry, 'Some Landmarks in the Progress of Industrial

Medicine', *Cambridge University Medical Society's Magazine*, vol. 12 (1935), pp. 63-72; E.R.A. Merewether, 'The British Tradition in Industrial Health', *British Journal of Industrial Medicine*, vol. 5 (1948), pp. 175-179; T. Ferguson, 'Early Scottish Essays in Industrial Health', *British Journal of Industrial Medicine*, vol. 5 (1948), pp. 180-184; A. Meiklejohn, 'Sixty Years of Industrial Medicine in Great Britain', *British Journal of Industrial Medicine*, vol. 13 (1956), pp. 155-162; L. Teleky, 'Lessons from the History of Lead Poisoning', *Industrial Medicine* (1940), pp. 17-20; T.A. Lloyd Davies, *The Practice of Industrial Medicine* (London, 1948), pp. 6-17; Lloyd Davies, 'Evolution of Concepts in Industrial Medicine', *British Journal of Industrial Medicine*, vol.23 (1966), pp. 165-172. For an overview see L.J. Goldwater, 'From Hippocrates to Ramazzini, Early History of Industrial Medicine', *Annals of Medical History*, vol. 8 (1936), pp. 27-35; A.R. Hall, 'Homo Fabricator: a New Species', *History and Philosophy of the Life Sciences*, vol. 2 (1980), pp. 193-214. On France see V.P. Comiti, 'Les maladies et le travail lors de la revolution industrielle française', *History and Philosophy of the Life Sciences*, vol. 2 (1980), pp. 215-239; A. Farge, 'Les artisans malades de leur travail', *Annales E.S.C.*, vol. 32 (1977), pp. 993-1006. On the United States see J.S. Felton, 'Industrial Medicine to Occupational Health and Safety: a 50-year Retrospective', *Occupational Health and Safety*, vol. 51 no. 9 (1982), pp. 14-22; B.J. Stern, *Medicine in Industry* (Commonwealth Fund, New York, 1946); C. Levenstein, 'A Brief History of Occupational Health in the United States', in B.S. Levy and D.H. Wegman, *Occupational Health* (Boston and Toronto, 1983), pp. 11-12; D.M. Berman, 'Why Work Kills. A Brief History of Occupational Health and Safety in the United States', *International Journal of Health Services*, vol. 7 (1977), pp. 63-87; Berman, *Death on the Job: Occupational Health and Safety Struggles in the United States* (New York and London, 1979); V. Navarro, *Medicine Under Capitalism* (New York and London, 1976); G. Rosen, 'Early Studies of Occupational Health in New York City in the 1870s', *American Journal of Public Health*, vol. 67 (1977), pp. 1100-1102. On Russia and the Soviet Union see B. Haigh, 'The Early Development of Industrial Medicine in Russia', MD thesis Cambridge, 1974; H.E. Sigerist, *Socialised Medicine in the Soviet Union* (London, 1937). On Ramazzini see, G. Rosen (ed.), *Bernardo Ramazzini, Diseases of Workers* (New York and London,

1964). For further references to Italy see chapter 13.

3   H.E. Sigerist, 'The Social History of Medicine', *The Western Journal of Surgery, Obstetrics and Gynaecology*, (1940), pp. 715-722; Sigerist, 'Historical Background of Industrial and Occupational Diseases', *Bulletin of the New York Academy of Medicine*, 2nd ser. vol. 12 (1936), pp. 597-609; Sigerist, *On the Sociology of Medicine* (New York, 1960).

4   T.M. Legge, 'Industrial Diseases in the Middle Ages', *Journal of Industrial Hygiene*, vol. 1 (1919-20), pp. 475-483; Legge, 'The Spirit of Work under the Craft Guilds of the Middle Ages', *Journal of Industrial Hygiene*, vol. 1 (1919-20), pp. 550-556; Legge, *Shaw Lectures on Thirty Years Experience of Industrial Maladies* (London, 1929); L. Teleky, *History of Factory and Mine Hygiene* (New York, 1948).

5   G. Rosen, 'On the Historical Investigation of Occupational Diseases. An Aperçu', *Bulletin of the History of Medicine*, vol. 5 (1937), pp. 941-946; Rosen, *Miners' Diseases;* Rosen, *From Medical Police to Social Medicine* (New York, 1974); J.A. Ryle, 'Social Medicine as a Discipline', *British Journal of Industrial Medicine*, vol. 2 (1945), pp. 108-110.

6   T. Ferguson, *The Dawn of Scottish Social Welfare* (London, 1948); Ferguson, *Scottish Social Welfare (1864-1914)* (Edinburgh, 1958); A. Meiklejohn, 'Doctor and Workman', *British Journal of Industrial Medicine*, vol. 7 (1950), pp. 105-116; Meiklejohn, 'Industrial Health — Meeting the Challenge', *British Journal of Industrial Medicine*, vol. 16 (1959), pp. 1-10; Meiklejohn, 'History of Lung Diseases of Coal Miners in Great Britain', *British Journal of Industrial Medicine*, vol. 8 (1951), pp. 127-137; vol. 9 (1952), pp. 93-98, 208-221; Meiklejohn, 'Health Hazards in the North Staffordshire Pottery Industry 1688-1945', *Journal of the Royal Sanitary Institute*, vol. 66 (1946), pp. 516-525; Meiklejohn, 'A House Surgeon's Observations on Bronchitis in North Staffordshire Pottery Workers in 1864', *British Journal of Industrial Medicine*, vol. 14 (1957), pp. 211-213; Meiklejohn, 'The Origin of the Term Pneumoconiosis', *British Journal of Industrial Medicine*, vol. 17 (1960), pp. 155-160; Meiklejohn, 'The Successful Prevention of Silicosis among China Biscuit Workers in the North Staffordshire Potteries', *British Journal of Industrial Medicine*, vol. 20 (1963), pp. 255-263; Meiklejohn, 'The History of Occupational Respiratory Disease in the North Staffordshire Pottery Industry', in C.N. Davies (ed.), *Health*

Conditions in the Ceramic Industry (Oxford, 1969), pp. 3-14; Meiklejohn, 'Memories and Reflections. The Wyers Memorial Lecture, 1957', *Transactions of the Association of Industrial Medical Officers*, vol. 8 (1958), pp. 44-57; R.S.F. Schilling, 'Andrew Meiklejohn [1899—1970] - a Profile in Occupational Health', *Transactions of the Association of Industrial Medical Officers*, vol. 9 (1959), pp. 79-80.

7   E.P. Thompson and E. Yeo (eds.), *The Unknown Mayhew. Selections from the Morning Chronicle* (Harmondsworth, 1973); A. Tilgher, *Work: What It has Meant to Men Through the Ages* (New York, 1930); A. Briggs, *Social Thought and Social Action: Seebohm Rowntree* (London, 1961); S. Webb and B. Webb, *The State and the Doctor* (London, 1910); J.L. Hammond and B. Hammond, *The Village Labourer* (London, 1911); Hammond and Hammond, *The Town Labourer* (London, 1917); A. Clark, *Working Life of Women in the Seventeenth Century* (London, 1919; reprinted London, 1982); I. Pinchbeck, *Women Workers and the Industrial Revolution 1750 - 1850* (1930), 3rd edn (London, 1981); B.L. Hutchins, *Women in Modern Industry* (London, 1915; reprinted East Ardsley, 1978); B.L. Hutchins and A. Harrison, *A History of Factory Legislation* (London, 1903); M.C. Buer, *Health, Wealth and Population in the Early Days of the Industrial Revolution 1760-1815* (London, 1926); R. Titmuss, 'Industrialization and the Family', in *Essays on the Welfare State*, 2nd edn (London, 1963), pp. 104-118.

8   W.H. Dawson, *Social Insurance in Germany 1883-1911* (New York, 1911); A. Shadwell, *Industrial Efficiency. A Comparative Study of Industrial Life in England, Germany and America* (London, 1909); W. Mommsen (ed.), *The Emergence of the Welfare State in Britain and Germany 1850-1950* (London, 1981); G.A. Ritter, *Sozialversicherung in Deutschland und England* (München, 1983).

9   D.V. Glass, *Numbering The People* (London and New York, 1978); F.E. Manuel and F.P. Manuel, *Utopian Thought in the Western World* (Oxford, 1979); R.H. Tawney, *Religion and the Rise of Capitalism* (London, 1926); M. Weber, *The Protestant Ethic and the Spirit of Capitalism* (London, 1930); F.D. Klingender, *Art and The Industrial Revolution* (London, 1947); note Klingender, *The Condition of Clerical Labour in Britain* (London, 1935); C. Webster, *The Great Instauration. Science, Medicine and Reform, 1626-1660* (London, 1975). For social interpretation of science and medicine see P. Corsi and P.J. Weindling (eds.), *Information*

Sources for the History of Science and Medicine (London, 1983).

10   T. Ferguson, 'Employment and Health', *Edinburgh Medical Journal*, vol. 53 (1945), pp. 252-261.

11   A. Nugent, 'Fit for Work: the Introduction of Physical Examinations in Industry', *Bulletin of the History of Medicine*, vol. 57 (1983), pp. 578-595; J. Colie, *Malingering and Feigned Sickness* (London, 1913); T. Ferguson, 'Early Scottish Essays in Industrial Health', *British Journal of Industrial Medicine*, vol. 5 (1948), pp. 180-184.

12   G. Weber, 'Science and Society in Nineteenth Century Anthropology', *History of Science*, vol. 12 (1974), pp. 260-283. On eugenics and public health see J. Lewis, *The Politics of Motherhood* (London and Montreal, 1980); G. Williams, 'Compulsory Sterilization of Welsh Miners, 1936', *Llafur*, vol. 3 (1981/2), pp. 67-73; P.J. Weindling, 'Die Preussische Medizinalverwaltung und die Rassenhygiene', *Zeitschrift für Sozialreform*, (1984), pp. 675-687. On women's work see A.M. Anderson, *Women in the Factory. An Administrative Adventure, 1893 to 1921* (London, 1922); E. Mappen, *Helping Women at Work. The Women's Industrial Health Council 1889-1914* (London, 1985); P. Summerfield, *Women Workers in the Second World War* (London, 1985); K. Hausen (ed.), *Frauen suchen ihre Geschichte* (München, 1983); J. Lewis, *Women in England 1870-1950* (Brighton and Bloomington, 1984).

13   J. Ettling, *Germ of Laziness. Rockefeller Philanthropy and Public Health in the New South* (Cambridge, Mass. and London, 1981); J.C. Garcia 'The Laziness Disease', *History and Philosophy of the Life Sciences*, vol. 3 (1981), pp. 31-59.

14   M. Maclean and M. Jeffreys, 'Disability and Deprivation' in D. Wedderburn (ed.), *Poverty, Inequality and Class Structure* (Cambridge, 1974), pp. 165-179; F.B. Smith, *The People's Health* (London and New York, 1979), pp. 170-173, 330-332; G.A. Peters and B.J. Peters, *Sourcebook on Asbestos Diseases* (New York and London, 1980); J.C. McVitie, 'Asbestosis in Great Britain', *Annals of the New York Academy of Sciences*, vol. 132 (1965), pp. 128-138; P. Brodeur, *Expendable American* (New York, 1979); K. Figlio, 'Chlorosis and Chronic Disease in Nineteenth-Century Britain. The Social Constitution of Somatic Illness in a Capitalist Society', *Social History*, vol. 3 (1978), pp. 167-179, and *International Journal of Health Services*, vol. 8 (1979), pp. 589-617; I.S.L. Loudon, 'Leg Ulcers in the Eighteenth and early Nineteenth Centuries', *The Journal of the Royal College of General*

Practitioners, vol. 31 (1981), pp. 263-273. On the history of industrial dermatology see R.O. Noojin, 'Brief History of Industrial Dermatology', *Archives of Dermatology and Syphilology,* vol. 70 (1954), pp. 723-731; G.A. Hodgson, 'The History of Coal Miners' Skin Diseases', in J. Cule (ed.), *Wales and Medicine* (Llandysul, 1975), pp. 54-67; Hodgson, 'Dermatology and History in Wales', *British Journal of Dermatology,* vol. 90 (1974), pp. 699-671. E. Williams, 'The History of Brucellosis in Wales', in Cule, *Wales and Medicine,* pp. 75-87. On deformities see G. Rosen, 'The Worker's Hand', *Ciba Symposium,* vol. 5 (1942-3), pp. 1307-1318; Rosen, 'The Miner's Elbow', *Bulletin of the History of Medicine,* vol. 8 (1940), pp. 1249-1251; T. Sommerfeld, *Atlas der Gewerblichen Gesundheitspflege* (Berlin, 1926). For a case-study of a disaster see L. Conti, *Visto da Seveso: l'evento straordinario e l'ordinario amministrazione* (Milano, 1977). On radiation see S.B. Dewing, *Modern Radiology in Historical Perspective* (Springfield, 1969); R.J. Reynolds, 'Sixty Years of Radiology', *British Journal of Radiology,* vol. 29 (1956), pp. 238-245; F.G. Spear, 'Radiation Martyrs', *British Journal of Radiology,* vol. 29 (1956), p. 273; *Radiological Hazards to Patients* (H.M.S.O., London, 1961); S. Berg, 'History of the First Survey on the Medical Effects of Radioactive Fall-out', *Military Medicine,* vol. 124 (1959), pp. 782-785; B. Wynne, *Rationality and Ritual. The Windscale Inquiry and Nuclear Decisions in Britain* (British Society for the History of Science, Halfpenny Furze, 1982); R. J. Cloutier, 'Florence Kelley and the Radium Painters', *Health Physics,* vol. 39 (1980), pp. 711-716.

15   T. Holman, 'Historical Relationship of Mining, Silicosis and Rock Removal', *British Journal of Industrial Medicine,* vol. 4 (1947), pp. 1-29; A. Batty Shaw, 'Knappers' Rot. Silicosis in East Anglian Flint Knappers', *Medical History,* vol. 25 (1981), pp. 151-168; E. Posner, 'Milestones in the History of Mineral Dust Pneumoconiosis', in J. Cule (ed.), *Wales and Medicine* (Llandysul, 1975), pp. 42-53; J. Gwynne Morgan, 'The Place of Nickel in the History of Industrial Diseases', in Cule, *Wales and Medicine,* pp. 68-74; P. Hugh-Jones and C.M. Fletcher, *The Social Consequences of Pneumoconiosis among Coalminers in South Wales* (H.M.S.O., London, 1951); J.T. Hart, 'The Health of Coal Mining Communities', *Journal of the Royal College of General Practitioners,* vol. 21 (1971), pp. 517-528; F. Liddell, 'Morbidity of British Coal Miners in 1961-1962', *British Journal of Industrial*

Medicine, vol. 30 (1973), pp. 15-24; G. Rees, 'Health, the Distribution of Health Services and Poverty in Wales', in G. Rees and T.L. Rees (eds.), *Poverty and Social Inequality in Wales* (London, 1980), pp. 93-115; Enid M. Williams, *The Health of Old and Retired Coalminers in South Wales* (Cardiff, 1933); A. Bryan, *The Evolution of Health and Safety in Mines* (Letchworth, 1975); L. Doyal and S.S. Epstein, *Cancer in Britain. The Politics of Prevention* (London and Sydney, 1983).

16   M.P. Johnson, 'Grinders' Asthma in Sheffield', *Bulletin of the Society for the Social History of Medicine*, no.16 (1975), pp. 7-8; P. Kinnersly, *The Hazards of Work: How to Fight Them* (Pluto Press, London, 1973), p. 16. For further workers' songs see A. Meiklejohn, 'The History of Occupational Respiratory Disease in the North Staffordshire Pottery Industry', in C.N. Davies (ed.), *Health Conditions in the Ceramic Industry* (Oxford, 1969), pp. 10-14.

17   W.G. Carson, *The Other Price of Britain's Oil* (Oxford, 1981). For a case-study in the politics of accidents see D.F. Crew, 'Steel, Sabotage and Socialism: The Strike at the Dortmund "Union" Steel Works in 1911', in R.J. Evans (ed.), *The German Working Class 1888-1933* (London and Totowa, 1982), pp. 103-141.

18   G. Rosen, 'Occupational Health Problems of English Painters and Varnishers in 1825', *British Journal of Industrial Medicine*, vol. 10 (1953), pp. 195-199.

19   H. Levy, 'The Economic History of Sickness and Welfare Benefit Before the Puritan Revolution', *Economic History Review*, vol. 13 (1943), pp. 42-57; Levy, 'The Economic History of Sickness and Medical Benefit Since the Puritan Revolution', *Economic History Review*, vol. 14 (1944-5), pp. 135-160.

20   A. Wilson and A. Levy, *Workmen's Compensation* (2 vols., Oxford, 1939); K. Figlio, 'How Does Illness Mediate Social Relations? Workmen's Compensation', in Wright and Treacher, *Social Construction;* G.B. Rooke, 'History of Compensation for Industrial Lung Disease', *Bulletin of the Society for the Social History of Medicine*, no.16 (1975), pp. 6-7; J. Benson, 'The Compensation of English Coal Miners and their Dependants for Industrial Accidents, 1860-1897', PhD Thesis, University of Leeds, 1974; D. Hanes, *The First British Workmen's Compensation Act, 1897* (New Haven and London, 1968); P. Bartrip, *The Wounded Soldiers of Industry* (Oxford, 1983).

21   D. Henschler, 'Exposure Limits. History, Philosophy, Future

Developments', *Annals of Occupational Hygiene*, vol. 28 (1984), pp. 79-92; T.M. Legge, 'Twenty Years' Experience of the Notification of Industrial Diseases', *Journal of Industrial Hygiene*, vol. 1 (1919-20), pp. 590-596.

22 M. Berg (ed.), *Technology and Toil in Nineteenth Century Britain* (London and Atlantic Highlands, 1979); Berg, *The Machinery Question and the Making of Political Economy 1815-1848* (Cambridge, 1980); A.P. Robson, *On Higher than Commercial Grounds: The Factory Controvosy, 1830-1853* (New York, 1985).

23 H.A. Waldron, 'A Brief History of Scrotal Cancer', *British Journal of Industrial Medicine*, vol. 40 (1983), pp. 340-401; J. Rule, *The Experience of Labour in Eighteenth-century Industry* (London, 1981).

24 J.V. Pickstone and S.V.F. Butler, 'The Politics of Medicine in Manchester, 1788-1792: Hospital Reform and Public Health Services in the Early Industrial City', *Medical History*, vol. 28 (1984), pp. 227-249; Meiklejohn, 'Outbreak of Fever in Radcliffe Cotton Mills, 1784', *British Journal of Industrial Medicine*, vol. 16 (1959), pp. 68-70; G. Rosen, 'John Ferriar's "Advice to the Poor" ', *Bulletin of the History of Medicine*, vol. 11 (1942), pp. 222-227. For further bibliography see J.V. Pickstone, *Health, Disease and Medicine in Lancashire* (UMIST, Manchester, 1980).

25 W.R. Lee, 'Emergence of Occupational Medicine in Victorian Times', *British Journal of Industrial Medicine*, vol. 4 (1947), pp. 1-29; M.E. Rose, 'The Doctor in the Industrial Revolution', *British Journal of Occupational Medicine*, vol. 23 (1971), pp. 22-26; A. Meiklejohn (ed.), *Charles Turner Thackrah, The Effects of Arts, Trades and Professions on Health and Longevity* (Edinburgh and London, 1957); M.W. Thomas, *The Early Factory Legislation* (Leigh-on-Sea, 1948); T.M. Legge, 'Charles Turner Thackrah: A Pioneer in Industrial Hygiene', *Journal of Industrial Hygiene*, vol. 1 (1919-20), pp. 578-581; G. Rosen, 'Charles Turner Thackrah in the Agitation for Factory Reform', *British Journal of Industrial Medicine*, vol. 10 (1953), pp. 285-287; A. Meiklejohn, 'John Darwall, M.D. (1796-1833) and 'Diseases of Artisans' ', *British Journal of Industrial Medicine*, vol. 13 (1956), pp. 143-151; E.H. Greenhow, *Papers Relating to the Sanitary State of the People of England, General Board of Health, London 1858. With an introduction by Fraser Brockington* (Farnborough, 1973); R. Lambert, *Sir John Simon 1816-1904 and English Social*

Administration (London, 1963), pp. 331-339. On Arlidge see Elliott Isaacson, *The Forgotten Physician* (np, nd); E. Posner, 'Thomas Arlidge (1822-99) and the Potteries', *British Journal of Industrial Medicine,* vol. 30 (1973), pp. 266-270. On anthrax among woolsorters see M. Bligh, *Dr. Eurich of Bradford* (London, 1960). On the GP as mine surgeon see D.V. Bloor, 'Surgeon to the Mines', *Bulletin of the Society for the Social History of Medicine,* no. 30-31 (1982), p. 8; A.D. Morris, 'Two Colliery Doctors. The Brothers Armstrong of Trehorchy', in Cule, *Wales and Medicine,* pp. 208-215.

26   A. Landsborough Thomson, *Half a Century of Medical Research* (HMSO, London, 1975), pp. 165-189; R. Schilling, 'Industrial Health Research. The Work of the Industrial Health Research Board 1918-44', *British Journal of Industrial Medicine,* vol. 1 (1944), pp. 145-152; Schilling, 'The Present Work of the Industrial Health Research Board of Great Britain', *Industrial Medicine,* vol. 15 (1946), pp. 125-127; H. M. Sinclair, 'Sherrington and Industrial Fatigue', *Notes and Records of the Royal Society of London* (1984), pp. 91-104.

27   S.A. Henry, *Cancer of the Scrotum in Relation to Occupation* (London, 1946); H.A. Waldron, 'A Brief History of Scrotal Cancer', *British Journal of Industrial Medicine,* vol. 40 (1983), pp. 390-401; J.R. Brown and J.L. Thornton, 'Percival Pott and Chimney Sweepers' Cancer of the Scrotum', *British Journal of Industrial Medicine,* vol. 14 (1957), pp. 63-70; Waldron, J.A.H. Waterhouse and N. Tesema, 'Scrotal Cancer in the West Midlands 1936-76', *British Journal of Industrial Medicine,* vol. 41 (1984), pp. 437-444; Doyal and Epstein, *Cancer;* W. Mass and C. Levenstein, 'Labor Relations, Technology, and Occupational Disease. Banning the Suck Shuttle in Massachusetts, 1911', Unpublished Paper Presented to the Business History Conference, Hartford, 1984; W.C. Hueper, *Occupational Tumours and Allied Diseases* (Springfield, 1942).

28   Examples of long-term studies are P.J. Taylor and J. Burridge, 'Trends in Death, Disablement, and Sickness in the British Post Office since 1891', *British Journal of Industrial Medicine,* vol. 39 (1982), pp. 1-10; M.L. Newhouse and H. Thompson, 'Epidemiology of Mesothelial Tumours in the London Area', *Annals of the New York Academy of Sciences,* vol. 132 (1965), pp. 579-588.

29   J.M. Eyler, *Victorian Social Medicine. The Ideas and Methods of*

William Farr (Baltimore and London, 1979); J. Cullen, *The Statistical Movement in Early Victorian Britain* (New York, 1975); M.A. Heasman, F.D.K. Liddell and D.D. Reid, 'The Accuracy of Occupational Vital Statistics', *British Journal of Industrial Medicine,* 2nd ser. vol. 15 (1958), pp. 141-146; M.R. Alderson, 'Some Sources of Error in British Occupational Mortality Data', *British Journal of Industrial Medicine,* vol. 29 (1972), pp. 245-254; C. Webster, *Health: Historical Issues* (Centre for Economic Policy Research, London, 1984); J. Alden, 'The Extent and Nature of Double Job Holding in Great Britain', *Industrial Relations Journal,* vol. 8 no. 3 (1977), pp. 14-30.

30   A.M. Anderson, 'Historical Sketch of the Development of Legislation for Injurious and Dangerous Industries in England', in T. Oliver (ed.), *Dangerous Trades* (London, 1902), pp. 24-43; A.E. Peacock, 'The Successful Prosecution of the Factory Acts 1833-1855', *Economic History Review,* 2nd ser. vol. 37 (1984), pp. 197-210; K.M. Lyell (ed.), *Memoir of Leonard Horner* (2 vols., Women's Printing Society, London, 1890); T.K. Djang, *Factory Inspection in Great Britain* (London, 1942); Her Majesty's Inspectors of Factories 1833-1983. *Essays to Commemorate 150 Years of Health and Safety Inspection* (HMSO, London, 1983); M.I. Roemer, 'From Factory Inspection to Adult Health Service', *British Journal of Industrial Medicine,* vol. 10 (1953), pp. 179-194; C. Nardinelli, 'Child Labour and the Factory Acts', *Journal of Economic History,* vol 40 (1980), pp. 739-756.

31   L. Teleky, 'Certifying Surgeons, a Century of Activity', *Bulletin of the History of Medicine,* vol. 16 (1944), pp. 382-388; S. Huzzard, The Role of the Certifying Surgeon in the State Regulation of Child Labour and Industrial Health 1833-1973, MA Thesis, University of Manchester, 1976; Huzzard, 'Certifying Surgeon: Child Labour and Industrial Disease', *Bulletin of the Society for the Social History of Medicine,* no. 16 (1975), pp. 5-6; J.A. Smiley, 'Some Aspects of the Early Evolution of the Appointed Factory Doctor Service', *British Journal of Industrial Medicine,* vol. 28 (1971), pp. 315-322; I. Pinchbeck and M. Hewitt, *Children in English Society* (2 vols., London and Toronto, 1969-1973); M. Cruickshank, *Children and Industry: Child Health and Welfare in North-West Teesside Towns During the Nineteenth Century* (Manchester, 1981); M.Kipling, *A Brief History of H.M. Medical Inspectorate* (Health and Safety Executive, London, 1979); W.R. Lee, 'Robert Baker: The First Doctor in the Factory

Department', *British Journal of Industrial Medicine,* vol. 21 (1964), pp. 85-93; T.M. Legge, 'Medical Supervision in Factories', *Journal of Industrial Hygiene,* vol. 2 (1920-21), pp. 66-71. On women inspectors see L. Walden, 'Profile on 'Bessie' Blackburn', *Occupational Safety and Health,* (August, 1983), pp. 22-23; F. Hasson, 'Women at Work', *Occupational Safety and Health,* (August, 1983), pp. 18-21.

32   F. Collier, 'Samuel Greg and Styal Mill', *Memoirs and Proceedings of the Manchester Literary and Philosophical Society,* vol. 85 (1941-3), pp. 139-154; R. Murray, 'Quarry Bank Mill', *British Journal of Industrial Medicine,* vol. 15 (1958), pp. 293-298; vol. 16 (1959), pp. 61-67; N. McKendrick, 'Josiah Wedgwood and Factory Discipline', *Historical Journal* , vol. 4 (1961), pp. 30-35. E. Posner, 'Eighteenth-century Health and Social Service in the Pottery Industry of North Staffordshire', *Medical History,* vol. 18 (1974), pp. 138-145; B. Davies, 'Empire and Identity: The 'Case' of Dr William Price', in D.B. Smith (ed.), *A People and a Proletariat* (London, 1980), pp. 72-93.

33   J.T. Carter, 'History of the Society of Occupational Medicine', *Journal of the Society of Occupational Medicine,* vol. 35 (1985); A.J. Amor and C. Sykes, 'The Training of the Industrial Nurse', *British Journal of Industrial Medicine,* vol. 1 (1944), pp. 81-89; A.A. Eagger, *Venture in Industry. The Slough Industrial Health Service 1947 - 1963* (London, 1965); 'The New Haven Industrial Medical Service', *The Connecticut State Medical Journal,* vol. 10 (1946), pp. 193-196; M. Jefferys and C.H. Wood, 'A Survey of Small Factories', *British Journal of Industrial Medicine,* vol. 77 (1960), pp. 10-24.

34   C. Fraser Brockington, *The Health of the Community* (London, 1954), pp. 222-232.

35   D.J. Rowe, *Lead Manufacturing in Britain. A History* (London and Canberra, 1983). On other environmental hazards see A.E. Dingle, 'The Worst Nuisance of All: Landowners, Alkali Manufacturers, and Air Pollution 1828-64', *Economic History Review,* 2nd ser. vol. 35 (1982), pp. 529-548.

36   P.H.J.H. Gosden, *The Friendly Societies in England 1815-1875* (Manchester, 1961); Gosden, *Self Help. Voluntary Associations in the 19th Century* (London, 1973); J. Burnett, D. Vincent and D. Mayall (eds.), *The Autobiography of the Working Classes* (Brighton, 1985); M.H. Marland, 'Medicine and Society in Wakefield and Huddersfield, 1780-1870', PhD thesis, Warwick, 1984, chapter

5; C.G. Hanson, 'Craft Unions, Welfare Benefits, and the Case for Trade Union Reform', *Economic History Review*, 2nd ser. vol. 28 (1975), pp. 243-259; J. Benson, 'English Coal-Miners' Trade-Union Accident Funds, 1850-1900', *Economic History Review*, 2nd ser. vol. 28 (1975), pp. 401-412; N. Whiteside, 'Welfare Insurance and Casual Labour: A Study of Administrative Intervention in Industrial Employment, 1906-26', *Economic History Review*, 2nd ser. vol. 32 (1979), pp. 507-522; R. Earwicker, 'A Study of the BMA - TUC Joint Committee on Medical Questions, 1935-1939', *Journal of Social Policy*, vol. 8 (1979), pp. 33 5-356; S.J. Woodall, *The Manor House Hospital* (London, 1966); Earwicker, 'Miners' Medical Services Before the First World War: the South Wales Coalfield', *Llafur*, vol. 3. no. 2 (1981), pp. 34-52; L. Bryder, 'Papworth Village Settlement — a Unique Experiment in the Treatment of the Tuberculous?', *Medical History*, vol. 28 (1984), pp. 372-390.

37   For references to Nazi Germany see the chapters by Labisch and Milles.

38   S. Haber, *Efficiency and Uplift: Scientific Management in the Progressive Era 1890-1920* (Chicago, 1964); L. Baritz, *The Servants of Power: A History of the Use of Social Science in American Industry* (New York, 1965); C.R. Littler, *The Development of the Labour Process in Capitalist Societies. A Comparative Study of the Transformation of Work Organization in Britain, Japan and the USA* (London, 1982); R. Price, 'The Labour Process and Labour History', *Social History*, vol. 8 (1983), pp. 57-75; S. Smith, 'Taylorism Rules OK? Bolshevism, Taylorism and the Technical Intelligentsia in the Soviet Union, 1917-41', *Radical Science Journal*, no. 13 (1983), pp. 3-27; C.S. Maier, 'Between Taylorism and Technocracy', *Journal of Contemporary History*, vol. 5 (1970), pp. 27-61; R.A. Brady, *The Rationalization Movement in German Industry* (Berkeley, 1938); H.Homburg, 'Anfänge des Taylor Systems in Deutschland vor dem ersten Weltkrieg. Eine Problemskizze unter besonderer Berücksichtigung der Arbeitskämpfe bei Bosch', *Geschichte und Gesellschaft*, vol. 4 (1978), pp. 170-195; H.J. Welch and C.S. Myers, *Ten Years of Industrial Psychology: an Account of the First Decade of the National Institute of Industrial Psychology* (London, 1932); R. Marriott, 'An Outline of the History and Work of the Medical Research Council's Industrial Psychology Research Group', *Occupational Psychology*, vol. 32 (1958), pp. 26-33.

39 E.P. Thompson, 'Time, Work Discipline and Industrial Capitalism', *Past and Present*, no. 38 (1967), pp. 56-97; M. Bienefeld, *Working Hours in British Industry: An Economic History* (London, 1972); B. Meakin, *Model Factories and Villages: Ideal Conditions of Labour and Housing* (London, 1905); S. Pollard, *The Genesis of Modern Management* (Harmondsworth, 1968); P. Joyce, *Work, Society and Politics. The Culture of the Factory in Later Victorian England* (London, 1980); H. Jones, 'Employers' Welfare Schemes in Inter-War Britain', *Business History*, vol. 25 (1983), pp. 61-75; J. Melling, 'Industrial Strife and Business Welfare Philosophy. The Case of the South Metropolitan Gas Company', *Business History*, vol. 21 (1979), pp. 163-179; Melling, ' "Non-Commissioned Officers": British Employers and their Supervisory Workers, 1880-1920', *Social History*, vol. 5 (1980), pp. 183-221; R. Heller, *Food for Work* (Sutcliffe Catering Group, nd); C. Williams Mitchell, *Dressed for the Job. The Story of Occupational Costume* (Poole, 1982); W. Ashworth, 'British Industrial Villages of the Nineteenth Century', *Economic History Review*, 2nd ser. vol. 3 (1950), pp. 378-387; J. Child, 'Quaker Employers and Industrial Relations', *Sociological Review*, vol. 12 (1964), pp. 293-315; K. Thomas, 'Work and Leisure in Pre-industrial Society: Conference Report', *Past and Present*, no. 29 (1984), pp. 50-66; A. Lüdtke, 'Arbeitsbeginn, Arbeitspausen, Arbeitsende', in G. Huck (ed.), *Sozialgeschichte der Freizeit* (Wuppertal, 1980), pp. 95-122; F. Herzberg, *Work and the Nature of Man* (London, 1968).

2 SOCIAL HISTORY OF OCCUPATIONAL MEDICINE AND OF FACTORY HEALTH SERVICES IN THE FEDERAL REPUBLIC OF GERMANY.

Alfons Labisch

I Social History of Medicine: a Definition and Theory

There is no consensus as to the definition and theory of a social history of medicine in the Federal Republic of Germany. The relevant issues, methods and theories derive from different social sciences, and are still keenly debated.[1] Yet even at such an early stage of development, the available material merits review. The first beginnings of a theory can then be formulated, and this can be applied, tested and revised.

Implicit in the many writings by the historical demographer, Arthur Imhof, is a categorisation of the social history of medicine.[2] Imhof has distinguished (a) biological events (e.g. climate); (b) biology as experienced (e.g. body and soul); (c) biology as influenced (e.g. healing and medicine); (d) biology as observed (e.g. fertility); (e) biology as administered (e.g. public health). Imhof's methodology was derived from the French *Annales* school renowned for its quantitative and demographically oriented approach. As to his research aims, Imhof's categories seem primarily descriptive, although he has continually been able to relate his research to current issues in health care.

I proposed an alternative typology in 1980. It was based on medical history, medical sociology, and the social history produced in the previous decade. I evaluated these diverse approaches with regard to their subject-matter, methods and empirical and theoretical aims.[3] I arrived at the following classification of the different types of social history of medicine, ranging from the empirical to the theoretical. Social history of medicine may be seen as (a) the history of the interaction of medicine with different social classes and groups, especially the relations of medicine with social movements; (b) the history of medical disciplines and the corresponding institutions which evolved from and responded to social developments; (c) the history of health and disease in a broad biological, social, economic, political and scientific context, using historical and sociological concepts and methods; (d) an historico-sociological analysis of health and disease and their social conditions within a chronological framework either for the purposes of systematic social science, or for systematic social reconstruction of the present out of the past with the aim of obtaining pragmatic, action-oriented (i.e. political) results; (e) or, finally, an historico-sociological analysis of health and disease (like d) aiming to develop or scrutinise hypotheses, models, or theories by deploying historical evidence. The social history of medicine can therefore be described as a discipline which covers a range of objects, methods and topics from empirically oriented history (a-c) to pragmatically and/or theoretically oriented sociology (d and e).

II Social History of Occupational Medicine in the Federal Republic of Germany: an Evaluation

If one takes a closer look at the second position (b) — which is primarily oriented to the social history of medicine, linked to public health and social medicine as well as to occupational medicine —, the following implications arise for occupational medicine. Firstly, occupational medicine and industrial medical services are seen as medical disciplines, which in turn derive from societal developments and influence society. Secondly, research on the history of occupational medicine *has* to be conceptualised as social history. It may be more empirically oriented (see above, positions a - c) or have more pragmatic and/or theoretical aims (positions d and e).

The development of occupational medicine, and related forms of preventive medicine and health care are very closely bound up with the development of especially dangerous forms of work. Given that labour processes evolve as part of broader social changes, the interrelation of the labour process with social conditions must be regarded as fundamental in determining conditions of occupational health. This can be illustrated by the ancient examples of the working rules and provisions for medical care for the builders of the Egyptian pyramids, and for the Sicilian mining slaves in the Roman Empire. The development of mines in the middle ages has been the painful breeding ground of theoretical and practical innovations, as with the prevention of accidents, of lung diseases, and of lead, mercury and carbon dioxide poisoning. The sixteenth-century studies by Agricola and Paracelsus show this.[4]

In contrast to the basic axiom of the interaction of occupational medicine and industrial health with economic and social conditions, occupational medicine is today in the Federal Republic dominated by scientific reductionism. A similar and obviously related reductionism prevails in social policy, in the legal processes for compensation and other occupationally related issues, and in medical services. Occupational medicine is primarily conceptualised as toxicology, physiology and mechanical engineering. Only recently have social epidemiology and social welfare law forced occupational medicine into a more ecological approach. The controversies have focused on the issue of illness as caused by work.[5]

A history of occupational medicine cannot exclude its social dimensions. But if a parallel is drawn with the history of medicine, we have to take note that there still survives an approach that only takes account of medicine in its narrowest sense. If this approach were to be adopted then either occupational medicine would be excluded altogether from medical history, or it would be reduced to the narrowest of issues associated with purely scientific developments, or to occasional contributions by those who are or have been professionally active in occupational medicine.

In the Federal Republic as in Switzerland and Austria, the history of occupational medicine has been written primarily by occupational physicians at the height of their careers. They become aware of the long term implications of their work. That this need not necessarily be a worthless exercise is shown by the works of the former Nestor of German occupational medicine, Franz Koelsch, who was a descendent (a great grand nephew) of the leading theoretician of medical police, Johann Peter Frank. In 1909 Koelsch became the first medical inspector of factories in Bavaria. In 1967 when in his nineties he wrote a comprehensive history of occupational medicine.[6] He recognised in his early years the importance of the historical dimensions of his work. In 1912 he published a monograph on Bernardino Ramazzini, whom he dubbed the father of occupational medicine, and he repeatedly stressed historical issues in other publications.[7] Another early historian of occupational medicine was the Austrian, Ludwig Teleky, who pioneered the development of social medicine as industrial medical officer in Düsseldorf. From 1921 until his flight from the Nazis, he often published on the history of occupational medicine.[8] The development of the industrial medical service has been described by Bauer and Mertens, and Merkert has written on medical theories of women's occupational diseases from Ramazzini to Hirt.[9]

It is fortunate when an active practitioner in occupational medicine receives an academic chair in the history of medicine. This happened when Heinrich Buess (who for many years was an industrial medical officer with the pharmaceutical manufacturers, CIBA) became professor of the history of medicine in Basel. His works show that practitioners of occupational medicine tend to interpret occupational medicine in its narrowest sense and exclude social and political dimensions of their subject. Yet in the copious quotations and comments which they make, it is impossible for them to adhere strictly to their reductionism.[10] If we add to Koelsch, Teleky and Buess, Edwin Rosner, Erna Lesky and the American George Rosen, who was profoundly influenced by the medical history of Leipzig in the early 1930s, we have a comprehensive view of the older German history of occupational medicine.[11] A recent example of a practising occupational physician, who has published historical researches relating to his work, is the miners' insurance doctor, Karl Boventer, who has also written more generally on miners' diseases and on the development of miners' sick funds.[12] Another example of a practitioner-historian is the director of the medical services of BASF, Alfred Michael Thiess.[13] Thiess' empirical approach is especially effective in bringing new material to light. He shows that over one hundred years before it became law in the Federal Republic that there should be a factory health service, there was in 1866 a factory doctor in the chemical industry. By the turn of the century factory doctors were able to provide comprehensive medical care for workers and their families, until this had to cease in 1929 owing to the monopolisation of outpatient treatment by doctors in panel practices. This was due to a ruling by the Reich Court of Appeal, which in 1929 banned treatment of a worker's dependants, and in the following years factory doctors were totally excluded from therapeutic medicine. Finally, as Gregor Kern has shown, the factory medical service was a means used by employers to attract a skilled workforce. The developing chemical industry was in especial need of a reliable and stable workforce.[14]

In addition to the above-mentioned authors and works, there are occasional pieces which only rarely rise above the anecdotal. These can be more revealing about current questions in occupational medicine than about the past. For example, Hans-Günther Häublein has complained that occupational medicine in the strongly production-oriented GDR is too geared to physiological and technical criteria. He argues against a narrow concept of occupational physiology. Recent articles on the history of accident prevention publicity critically contrast the breadth of earlier approaches with technocratic modern methods.[15]

This survey of the older literature demonstrates that the history of occupational medicine can only be worthwhile when it takes account of social and economic conditions associated with hazards at work. The corresponding scientific and medical contributions and the transformation of these into private and public counter-measures to improve conditions provide a contrast to prevailing economic and social conditions.

In the following section more recent, post-1970 literature will be considered with regard to the typology of the social history of medicine, outlined above.

III Occupational Medicine and Social Protest

State social policy has often been formulated in response to workers' demands for improved health and working conditions. The origins of many political, legal and welfare reforms can be understood as deriving from pressures of the labour movement.

Among works by the Leipzig (and from 1981 Magdeburg) medical historian, Gerd Moschke, deserving attention, are his studies of workers' reactions to the doctors' strike of 1904, and of the recognition of disease as a risk factor in equivalent terms to those of accidents.[16] The Leipzig medical historian, Karl-Heinz Karbe, has researched on demands by radical doctors for the improvement of working conditions and their influence on preventive legislation.[17] Karbe has studied the Berlin Poor Law doctor, Salomon Neumann, who took a major role in the 'Medical Reform' movement of the 1840s, but clashed with another leader of the movement, Rudolf Virchow, over the issue of workers' democratic control of medical facilities. In 1849 Neumann founded in Berlin a 'Health Care Association', that aimed to provide a lasting form of co-operative medical care based on continuous contact of doctors and workers. It also aimed to collect facts on health risk, to improve preventive medicine and to provide free treatment of the sick. The workers carried out their own epidemiological surveys.[18] This Health Care Association had a remarkably large membership of 9000. It was banned in 1853 as a political association, showing that when health is demanded on the basis of equality to improve conditions, its social potential is such as to provoke restrictive and repressive reaction.

In the Federal Republic the development of social policy, especially in the area of social insurance, has become a special object of historical research. The problems of occupational medicine and of factory health services amount to only part of the broader field of social policy. Without doubt social policy has had an enormous influence on the development of occupational medicine. For example, the original concept of the causal relationship between work and disease was drawn from the Industrial Injury Law of 1884, and was only later supplemented by monocausal, scientifically proven aetiologies of disease. There have been a number of contributions to the social history of factory health. Harro Jenss has traced the development of youth employment law.[19] Wolfgang Bocks has described how the Baden Factory Inspectorate took a pioneering initiative in employing women factory inspectors.[20] Ute Frevert has made an impressive study of workers' health insurance. She has discovered the massive development of associations akin to friendly societies prior to Bismarck's social insurance legislation.[21] Many contributions have been made by the Kassel welfare lawyer, Florian Tennstedt, who first dealt with the development of industrial injury compensation, and then investigated the social history of social insurance, as well as undertaking more theoretical analyses of social protest and social welfare.[22] Finally, two further works deserve mention: by Joachim Hohmann on the Occupational Diseases Statute of 12 May 1925,[23] and by Peter Hinrichs on industrial psychology in Germany from 1871 -1945.[24]

IV Occupational Medicine as Social History

When the third type of a social history of occupational medicine is carried out from broader historical and sociological perspectives, then a more generalised type of social history is attained. This status has been achieved by certain dissertations on the development of preventive medicine in the chemical industry, which because of its often dangerous technological innovations has been regarded as the classic breeding ground of occupational medicine since its beginnings. An example is the research by Kern, whose study of Hoechst concludes with an analysis of the significance of health measures for management and labour.[25] There is an interesting thesis by Breuer on the occupational health of musicians.[26] Otherwise in the Federal Republic there are only minor occasional pieces like lectures and introductions to new journals.[27] An example is an informative item by Schadewaldt.[28] A less than exemplary piece was produced by Jetter, who geared his survey to the founding of a new journal of occupational medicine, and after reaching the eighteenth century switched to the archetectural history of hospitals.[29] In the older literature on social history, Jürgen Kuczinsky's monumental works are a useful quarry for material. Valuable matter can also be sifted out from the older medical statistical, social hygienic and epidemiological standard works. Examples of such key texts are those by Friedrich Prinzing, Mosse and Tugendreich, Alfons Fischer and Alfred Grotjahn.[30] More recently, economic and social historians have become interested in the history of occupational medicine, and especially in the study of risks at work.[31] But occupational medicine and factory health have only just begun to arouse the attention of social historians, and only now has systematic historical research begun.

The situation in the German Democratic Republic is very different. Here too there are regional studies, as by the Halle medical historian, Wolfram Kaiser.[32] Studies of artisan and industrial working conditions, written in the vein of *Heimatkunde*, are also produced.[33] The gynaecologist and historian, Peter Schneck, has written about women workers in the Saxon Oberlausitz.[34] Penetrating analyses by the leading historian of public health, Dietrich Tutzke, also merit attention, as for example his account of the mirror industry.[35] Tutzke's case study is a model of how industrial processes, health risks, medical observations and socio-political responses can be historically integrated.

Under the directorship of Stanislaw Schwann, the Karl Sudhoff Institute for the History of Medicine and Science (the first ever such institute, founded in 1906) has made a major contribution to the history of occupational medicine and industrial health. The Institute made the history of industrial health a prority area of research. Besides the work of Gerd Moschke and Heinz Pilz, [36] the output of Karl-Heinz Karbe merits special consideration. In numerous publications he has succeeded in uniting medical, scientific, economic and political aspects into a unified social historical picture.[37] An example of one of his many major contributions to the history of nineteenth-century medicine is his research on phosphor necrosis of the jaw. He describes the changing scientific theories in relation to working conditions, before turning to medical theories of preventive legislation, and to the differing regional responses of medical officials. He has evaluated the extent to which the proposals were implemented and their effectiveness in reducing the disease. He has also made a general study of occupational medicine from 1780-1850, producing a text which is absolutely fundamental to the social history of industrial health in this period. He recently published a study of Salomon Neumann's attempts to improve workers' health in Berlin.

If a criticism is to be ventured of such a recent but highly important area of research as is the history of medicine in the GDR, it would be that the research uses primarily medical literature, and until recently has not exploited other methods and sources, as, for example, quantitative methods, or those drawn from the now burgeoning social history of everyday life. The research has adopted 1870 as an arbitrary cut-off point in time, after which no systematic research has been attempted. The Leipzig Institute has transferred its main focus of research to the history of psychiatry, owing to the change in Director from Schwann to Achim Thom.[38]

A new area of socio-historical research, which has emerged in both the FRG and GDR, is medicine under national socialism. A general work by Walter Wuttke-Groneberg has summed up Nazi medicine as 'achievement, extermination and distortion' *(Leistung, Vernichtung und Verwertung)*. Much general literature on these themes has dealt with problems of occupational medicine and factory health under the Nazis. In the ideology of achievement, occupational medicine was especially important.[39] Graesner has further investigated the implications of *Leistungsmedizin* in Nazi Germany.[40] Roth has undertaken a study of the drug *Pervitin* and of how it was associated with the ideology of achievement.[41] Wuttke-Groneberg has dealt more generally with ideological themes in medicine of the period.[42] Karbe has analysed the factory doctor system and occupational medicine in fascist Germany.[43] Moschke has considered attitudes in the Ministry of Labour *(Reichsarbeitsministerium)* to occupational diseases.[44]

V Social History of Medicine with a Pragmatic Aim

The fourth category or level of a social history of occupational medicine aims at systematic reconstruction of the present from the past and at obtaining pragmatic results. This has only been a recent characteristic of research projects in the FRG. Projects are all linked to the current discussion outlined above of the aims and state of occupational medicine. The Bremen project, 'Shattered Alternatives in Medicine', shows that current debates can be traced to their historical roots.[45] Broad ecological concepts of workers' illnesses and innovative methods for comprehensive health care can be found during early phases of industrialisation. An example is the Berlin Health Care Association, which was banned as a political group. In this sense those working on the social history of occupational medicine in Bremen have a pragmatic aim of broadening current scientific and socio-political discussions.

For the first time there is a research project that aims to document the relevant sources for a history of occupational medicine and industrial health in Germany from 1835.[46] It is supported by the *Bundesanstalt für Arbeitsschutz und Unfallversicherung*, Dortmund, a federal government office with responsibility for health and safety research and advice on legislation. The project employs the concept of 'dethematisation' as the basis for empirical studies. This concept does not ask when current issues had their origins and were 'thematised', but the reverse in that it asks when and why certain themes, concepts and theories of occupational medicine and preventive measures disappeared from the scientific discourse and from practice, and when these were 'de-thematised'. Dethematisation is therefore to be understood in terms of public indifference, professional reductionism and socio-political repression. First results by Rainer Müller, Dietrich Milles and Lothar Machtan are published here.[47] Three volumes of bibliographical and source materials have been prepared. The first volume consists of a bibliography of 5000 titles on the social origins of occupational medicine. The second volume contains sources on the social history of workers' diseases in selected occupations. The third volume includes sources on the history of occupational medicine. The source materials are available in a special archive at the University of Bremen, containing collections of personal papers of leading figures in occupational medicine.

VI Social History of Medicine with a Theoretical Aim

Finally, one of the few examples of a social history of occupational medicine with regard to its systematic sociological aspects should be mentioned. This is the work of the Düsseldorf medical sociologist, Christian von Ferber, who in other sociological works has consistently paid attention to historical dimensions. He has considered the problem of preventive medicine at work with regard to the pathogenicity of social conditions as a result of social change. In this he makes use of the theories of Norbert Elias' treatise on the *Civilizing Process*, with regard to the dependence of communities on conflicts between social groups in the course of historical change.[48]

VII Conclusions

Classification of recent works on medical history and medical sociology provides the basis for a theory of the social history of medicine. This shows that medicine past and present cannot neglect social factors. It is especially true for occupational medicine. I have reviewed historical works in German in the light of these theories, which suggest that the history of occupational health requires consideration of not only specific hazards at the workplace, but also of the underlying social and economic conditions. It is axiomatic that there has been the interaction of occupational medicine, factory health, the labour process and socio-economic circumstances. Currently, however, occupational medicine in the FRG is dominated by a scientific reductionism that is closely bound up with developments in social policy. The socio-legal paradigm of a causal connection between work and disease and the medical paradigm of a monocausal scientific aetiology go hand in hand. Only very recently have attempts been made to develop a broader and more ecological concept of health care at work. For the most part these experiments have historical precursors.

As a result of the contradiction between the reductionist paradigm and the need to take social factors into account, the history of occupational medicine has been haphazard and without sustained research on particular problems or without theoretical concepts. Practitioners of occupational medicine have limited their attention to the history of their discipline. Only very recently have social historians become interested in the connections between work and health.

For some time the only systematic research in the history of occupational health has been at the Karl Sudhoff Institute for the History of Medicine and Science at Leipzig in the GDR. The works of

Karl-Heinz Karbe merit attention as uniting medical, scientific, economic and political aspects into a historically unified picture. But unfortunately the research programme has been discontinued.

In contrast, there are in the FRG new efforts to produce a pragmatically oriented social history of occupational health. These efforts are closely associated with attempts to use epidemiological and ecological models, and to construct from these a broadened concept of occupational medicine and industrial health. These social historical works (examples of which are in the following chapters) are firmly rooted in hitherto unused source materials. They are guided by the concept of 'dethematisation': i.e. why and when certain theories of occupational medicine and factory health disappeared from science, politics and practice.

This chapter is dedicated to Hans Joachim Einbrodt, Director of the Institute of Hygiene and Occupational Medicine of the RWTH Aachen.

I wish to thank Karl-Heinz Karbe, Dietrich Milles and my translator, Paul Weindling, for their stimulating comments and suggestions. The investigation has been supported by the Fachbereich 4 of the Gesamthochschule Kassel.

Notes

1   Alfons Labisch and Reinhard Spree, 'Neuere Ergebnisse und Entwicklungen einer Sozialgeschichte der Medizin und des Gesundheitswesens in Deutschland im 19. und 20. Jahrhundert', *Berichte zur Wissenschaftsgeschichte*, vol. 5 (1982), pp. 209-223; Alfons Labisch, Reinhard Spree and Paul Weindling, 'Social History of Nineteenth- and Twentieth-Century German Medicine', *Bulletin of the Society for the Social History of Medicine*, no. 32 (1983), pp. 40-45.
2   Arthur E.Imhof (ed.), *Biologie des Menschen in der Geschichte. Beiträge zur Sozialgeschichte der Neuzeit aus Frankreich und Skandinavien*, (Kultur und Gesellschaft, vol. 3), (Stuttgart, 1978), pp. 13-78. For comprehensive bibliography, see pp. 372-402.

3   Alfons Labisch, 'Sozialgeschichte der Medizin. Methodolog-
    ische Überlegungen und Forschungsbericht', *Archiv für
    Sozialgeschichte*, vol. 20 (1980), pp. 431-469, see 449 f.

4   Heinrich Buess, 'Paracelsus und Agricola als Pioniere der
    Sozial- und Arbeitsmedizin', in Erna Lesky (ed.), *Sozialmedizin.
    Entwicklung und Selbstverständnis* (Darmstadt, 1977), pp. 7-25.

5   Regarding this fundamental controversy, see: Deutsche
    Forschungsgemeinschaft   (ed.),   *Denkschrift   zur   Lage   der
    Arbeitsmedizin und der Ergonomie in der Bundesrepublik Deutschland*
    (Boppard, 1980), and the critique of this by Gine Elsner et. al.,
    'Arbeitsmedizin und Ergonomie in der BRD. Stellungnahme zu
    einer Denkschrift der Deutschen Forschungsgemeinschaft',
    Jahrbuch für kritische Medizin (Argument-Sonderhefte 73), vol. 7
    (1981), pp. 167-178. For examples of other strategies in
    occupational health, see: Matthias Schmidt et. al. (eds.), *Arbeit
    und Gesundheitsgefährdung. Materialien zur Entstehung und
    Bewältigung arbeitsbedingter Erkrankungen*, (SWI Studienhefte 2),
    (Frankfurt/M., 1982), and the indirect reply of traditional
    occupational medicine: Joseph Rutenfranz, 'Arbeitsbedingte
    Erkrankungen — Überlegungen aus arbeitsmedizinischer
    Sicht', *Arbeitsmedizin, Sozialmedizin, Präventivmedizin*, vol. 11
    (1983), pp. 257-267. For a comprehensive survey of recent
    ecological concepts see Werner Maschewsky, 'Zum Stand der
    Belastungs- und Beanspruchungsforschung', *Soziale Welt*, vol.
    33 (1982), pp. 328-345.

6   Franz Koelsch, *Beiträge zur Geschichte der Arbeitsmedizin*,
    (Schriftenreihe der Bayerischen Landesärztekammer vol. 8),
    (n.p. [München], 1967).

7   Franz Koelsch, *Bernardo Ramazzini, der Vater der Gewerbehygiene*
    (Stuttgart, 1912); Koelsch, 'Die Erforschung der Berufskrank-
    heiten von der Wende des 18./19. Jahrhunderts bis zur
    Gegenwart', in Ernst W. Baader (ed.), *Handbuch der gesamten
    Arbeitsmedizin*, (Berlin, 1961), vol. 2 pt. 1, pp. 37-68. Baader,
    Lehrbuch der Arbeitsmedizin, vol. 1, 4th edn (Stuttgart, 1983), pp.
    3-11; Baader, 'Johann Peter Frank — seine Bedeutung für die
    Arbeitsmedizin', *Deutsches Ärzteblatt*, vol. 63 (1966), pp. 1735-
    1737, 1777-1781, 1812-1813, 1885-1887.

8   Ludwig Teleky, 'Geschichte der Erkenntnis der Staublunge in
    Deutschland', *Archiv für Gewerbepathologie*, vol. 3 (1932), pp. 418-
    470; Teleky, *History of Factory and Mine Hygiene* (New York
    1948), containing a comprehensive bibliography, pp. 285-317.

On Teleky see: Heinrich Buess, 'Der Wiener Sozialmediziner Ludwig Teleky (1872-1957) und sein 'History of factory and mine hygiene' ', *Wiener medizinische Wochenschrift*, vol. 131 (1981), pp. 479-483; Esteban Rodriguez Ocana, 'Aproximacion al concepto practico de la Medicina Social en Ludwig Teleky (1872-1957)', *Dynamis*, vol. 2 (1982), pp. 299-323.

9 On the development of the industrial medical service, see: M.Bauer, 'Der gewerbeärztlicher Dienst. Ein Blick über 50 Jahre', *Bundesarbeitsblatt*, (1954), Sonderheft '100 Jahre Gewerbeaufsicht', pp. 29-32; also, A.Mertens, '125 Jahre Gewerbeaufsicht', *Zentralblatt für die Arbeitsmedizin*, vol. 28 (1978), pp. 144-145; Brigitte Merkert, Berufskrankheiten der Frauen als Problem der Medizin von Bernardino Ramazzini (1718) bis Ludwig Hirt (1873), med. diss., Mainz 1983.

10 Heinrich Buess, 'Die Erforschung der Berufskrankheiten bis zum Beginn des industriellen Zeitalters', in Baader, *Handbuch der gesamten Arbeitsmedizin*, vol. 2/1, pp. 15-36. Buess, 'Über den Beitrag deutscher Ärzte zur Arbeitsmedizin des 19. Jahrhunderts', (Studien zur Medizingeschichte des 19. Jahrhunderts 1), (Stuttgart, 1967), pp. 166-178. For a bibliography of Buess, see Lesky (ed.), *Sozialmedizin*, pp. 463-464.

11 Edwin Rosner, 'Die Bedeutung des Annaberger Stadtarztes Martin Pansa für die Geschichte der Gewerbehygiene', *Sudhoffs Archiv*, vol. 37 (1953), pp. 357-361; Rosner, 'Ulrich Ellenbog und die Anfänge der Gewerbehygiene', *Sudhoffs Archiv*, vol. 38 (1954), pp. 104-110; Rosner, 'Die Berufskrankheiten in der Predigtliteratur des 16. Jahrhunderts', *Sudhoffs Archiv*, vol. 41 (1957), pp. 193-206; Rosner, 'Hohenheims Bergsucht Monographie', *Medizinhistorisches Journal*, vol. 16 (1981), pp. 20-52; Erna Lesky, *Arbeitsmedizin im 18. Jahrhundert. Werksarzt und Arbeiter im Quecksilberbergwerk Idria* (Wien, 1956); Lesky, 'Ulrich Ellenbog und sein arbeitsmedizinsches Merkblatt, Montfort', vol. 3-4 (1972), pp. 612-618; George Rosen, *The History of Miner's Disease* (New York, 1943); Rosen, *A History of Public Health* (New York, 1958). For a bibliography of Rosen see: Charles E.Rosenberg (ed.), *Healing and History: Essays for George Rosen* (Dawson, 1979), pp. 252-262.

12 Karl Boventer, 'Zur Geschichte der Knappschaftsärzte im Steinkohlenbergbau', *Sudhoffs Archiv*, vol. 48 (1964), pp. 54-62; Boventer, 'Zur Geschichte der Bergbaumedizin und

Knappschaft im Steinkohlenbergbau an der Wurm', *Zeitschrift des Aachener Geschichtsvereins*, vol. 78 (1967), pp. 210-256.

13 Alfred Michael Thiess and Hans Dieter Flach, 'Über die Pioniertätigkeit der ersten Werksärzte Deutschlands', *Zentralblatt für Arbeitsmedizin*, vol. 20 (1970), pp. 81-87; Thiess, 'Werksärztliche Aufgaben der BASF 1866-1972', *Arbeitsmedizin, Sozialmedizin, Arbeitshygiene*, vol. 7 (1972), pp. 105-110.

14 Bernd Flohr, *Arbeiter nach Mass. Die Disziplinierung der Fabrikarbeiterschaft während der Industrialisierung Deutschlands im Spiegel von Arbeitsordnungen*, (Campus Forschung 221) (Frankfurt/M., 1981). For Kern see note 25.

15 Hans-Guenther Häublein, 'Einige Lehren der Vergangenheit für die Entwicklung der Arbeitsmedizin in der Deutschen Demokratischen Republik', *Zeitschrift für die gesamte Hygiene*, vol. 18 (1972), pp. 328-331. See also U. Renker, 'Zur Entwicklung des Betriebsgesundheitswesens in der DDR', *Wissenschaftliche Zeitschrift der Martin Luther Universität Halle - Wittenberg*, mathemathische-naturwissenschaftliche Reihe, vol. 25 (1976), pp. 95-98. For recent articles in the FRG, see: Oskar Lang, '75 Jahre Ausstellung von Arbeitsschutzeinrichtungen in München', *Zentralblatt für Arbeitsmedizin*, vol. 25 (1975), pp. 268-271; Anon., 'Die erste Ausstellung für Unfallverhütung in Berlin', *Die Berufsgenossenschaft*, vol. 27 (1975), pp. 379-383; Hans-Joachim Symanski, '40 Jahre Arbeitsmedizin, wie ich sie erlebte. Ein Rückblick und Ausblick', *Therapie der Gegenwart*, vol. 114 (1975), pp. 1545-1565.

16 Gerd Moschke, 'Der Kampf der Leipziger Arbeiter um eine neue Organisation der medizinischen Betreuung im Ärztestreik 1904', *Zeitschrift für ärztliche Fortbildung*, vol. 69 (1975), pp. 1141-1143; Moschke, 'Voraussetzungen der ersten gesetzlichen Gleichstellung der berufsbedingten Erkrankungen mit den Betriebsunfällen in Deutschland', *Zeitschrift für die gesamte Hygiene*, vol. 23 (1977), pp. 928-930.

17 Karl-Heinz Karbe, 'Über Forderungen deutscher Ärzte zur Verbesserung der Gesundheitsverhältnisse der Fabrikarbeiter im Jahrzehnt der bürgerlichen Revolution von 1848', *NTM*, vol. 8 (1971), pp. 45-53; Karbe, 'Zur Entstehungsgeschichte der ersten preussischen Verordnung über 'sanitätspolizeiliche Einrichtungen der Zündwarenfabriken' vom 29. Oktober 1857', *NTM*, vol. 16 (1979), pp. 83-94.

18 Karl-Heinz Karbe, 'Zur Geschichte des Berliner Gesundheitsp-

flegeverein der deutschen Arbeiterverbrüderung', *Deutsches Gesundheitswesen*, vol. 28 (1973), pp. 1621-1625, 2204-2208; Karbe, 'Die Berichte Salomon Neumanns über den Gesundheitspflegeverein der Berliner Arbeiterverbrüderung und den Berliner Gesundheitspflegeverein (1849-1853)', *Wissenschaftliche Zeitschrift der Martin Luther Universität Halle - Wittenberg*, mathemathische-naturwissenschaftliche Reihe, vol. 23 no. 4 (1974), pp. 66-72; Karbe, *Salomon Neumann. 1819-1908. Wegbereiter sozialmedizinischen Denkens und Handelns. Ausgewählte Texte*, (Sudhoffs Klassiker der Medizin, new series 3) (Leipzig, 1983). See also note 35.

19   Harro Jenss, 'Werksärztliche Versorgung in der BRD', *Argument*, no. 78 (1973), pp. 9-51; Jenss, Zur Entwicklung des Jugendarbeitsschutzes von den Anfängen der Industrialisierung bis zur Gegenwart, med. Diss., Frankfurt/M., 1977.

20   Wolfgang Bocks, *Die Badische Fabrikinspektion. Arbeiterschutz, Arbeiterverhältnisse und Arbeiterbewegung in Baden 1879 bis 1914*, (Forschungen zur oberrheinischen Landesgeschichte, vol. 27) (Freiburg/B., 1978).

21   Ute Frevert, 'Arbeiterkrankheit und Arbeiterkrankenkassen im Industrialisierungsprozess Preussens (1840-1870)', in Werner Conze and Ulrich Engelhardt (eds.), *Arbeiterexistenz im 19. Jahrhundert: Lebensstandard und Lebensgestaltung deutscher Arbeiter und Handwerker* (Stuttgart, 1981), pp. 293-319; Frevert, *Krankheit als politisches Problem 1770 - 1880. Soziale Unterschichten im Preussen zwischen medizinischer Polizei und staatlicher Sozialversicherung* (Göttingen, 1984).

22   Florian Tennstedt, *Berufsunfähigkeit im Sozialrecht. Ein soziologischer Beitrag zur Entwicklung der Berufsunfähigkeitsrenten in Deutschland* (Frankfurt/M., 1972). Tennstedt, 'Sozialgeschichte der Sozialversicherung', in M.Blohmke et. al. (eds.), *Handbuch der Sozialmedizin*, vol. 3 (Stuttgart, 1976), pp. 385-492; Tennstedt, *Sozialgeschichte der Sozialpolitik in Deutschland vom 18. Jahrhundert bis zum 1. Weltkrieg* (Göttingen, 1981); Tennstedt, *Vom Proleten zum Industriearbeiter. Arbeiterbewegung und Sozialpolitik in Deutschland 1800 - 1914* (Köln, 1983).

23   Joachim S. Hohmann, *Berufskrankheiten und Unfallversicherung. Vorgeschichte und Entstehung der ersten Berufskrankheiten-Verordnung vom 12. Mai 1925. Ein Beitrag zum 100jährigen Bestehen der deutschen Unfallgesetzgebung* (Köln 1983).

24 Peter Hinrichs, *Um die Seele des Arbeiters. Arbeitspsychologie, Industrie- und Betriebssoziologie in Deutschland. 1871-1945* (Köln 1981).

25 Gregor Kern, Der Beginn werksärztlicher Dienste und betrieblicher Sozialeinrichtungen in der Chemischen Industrie. Am Beispiel der Farbwerke Hoechst AG vorm. Meister Lucius und Bruening, med. Diss., Heidelberg, 1973. Markus Hämmerle, *Die Anfänge der Basler Chemischen Industrie im Lichte von Arbeitsmedizin und Umweltschutz,* (Basler Veröffentlichungen zur Geschichte der Medizin und Biologie, vol. 32) (Basel, 1979).

26 Rudolf Breuer, Berufskrankheiten von Instrumentalmusikern aus medizinhistorischer Sicht (vom 15. Jahrhundert bis 1930), med. Diss., Mainz, 1982.

27 For example, Erich Posner, 'Zur Geschichte der Staublunge. Nach zwei Vorträgen an der Universität Zürich, Juni 1974', *Gesnerus,* vol. 33 (1976), pp. 48-64.

28 Hans Schadewaldt, 'Arbeitsmedizin — Geschichte und Ausblick', *Medizinische Welt,* vol. 25 (1974), pp. 386-393; Schadewaldt, 'Vom Comptoir zum Grossraumbüro', *Zentralblatt fur Bakteriologie und Hygiene,* I. Abteilung. Orig. B vol. 158, (1973), pp. 287-304.

29 Dieter Jetter, 'Vorläufer und Ansätze zur späteren Arbeitsmedizin (16. Jahrhundert). Geschichte der Arbeitsmedizin, Teil I', *Arbeitsmedizin - Sozialmedizin - Sozialhygiene,* vol. 1 (1966), pp. 162-165. Jetter, 'Grundlagen und Entwicklung der Arbeitsmedizin im 17. Jahrhundert. Geschichte der Arbeitsmedizin, Teil II', *Arbeitsmedizin - Sozialmedizin - Sozialhygiene,* vol. 1 (1966), pp. 206-209; Jetter, 'Bernardino Ramazzini und die Arbeitsmedizin des 18. Jahrhunderts. Geschichte der Arbeitsmedizin, Teil III', *Arbeitsmedizin - Sozialmedizin - Sozialhygiene,* vol. 1 (1966), pp. 243-248. See also Jetter's contributions to the history of hospitals: Jetter, *Grundzüge der Hospitalgeschichte* (Darmstadt, 1973). Jetter, *Grundzüge der Krankenhausgeschichte* (1800 - 1900) (Darmstadt, 1977). For a critique of the archetectural approach, see Alfons Labisch, 'Krankenhauspolitik und Krankenhausgeschichte', *Historia Hospitalium,* vol. 13 (1980), pp. 217-233 and Labisch, 'Das Krankenhaus in der Gesundheitspolitik der deutschen Sozialdemokratie vor dem ersten Weltkrieg', *Medizinische Soziologie. Jahrbuch,* vol. 1 (1981), pp. 126-151; M. Stollenwerk, 'Die bauliche Entwicklung der Krankenversorgung der

Industriestadt Stolberg im Rheinland vor dem Hintergrund der Ursprunge der Messingindustrie', *Historia Hospitalium*, vol. 13 (1979-1980), pp. 164-176.

30   Jürgen Kuczinsky, *Die Geschichte der Lage der Arbeiter unter dem Kapitalismus*, 38 vols. (Berlin-Ost, 1960), see especially vol. 4 (1967), pp. 385-415, and vol. 19 (1968), pp. 200-209; Friedrich Prinzing, *Handbuch der medizinischen Statistik* (Jena, 1906), 2nd edn (1931); M. Mosse and G. Tugendreich (eds.), *Krankheit und soziale Lage* (München, 1913), reprinted (Göttingen, 1977); Alfons Fischer, *Grundriss der sozialen Hygiene*, 2nd edn (Karlsruhe, 1925); Alfred Grotjahn, *Soziale Pathologie* (Berlin, 1912), 2nd edn (1915), 3rd edn (1923), reprinted ([West-]Berlin, 1977).

31   Dirk Blasius, ' 'Volksseuchen': Zur historischen Dimension von Berufskrankheiten', *SOWI. Sozialwissenschaftliche Information für Unterricht und Studium*, vol. 6 (1977), pp. 55-60; Herbert Aagard, 'Gefahren und Schutz am Arbeitsplatz in historischer Perspektive. Am Beispiel des Nadelschleifens und Spiegelbelegens im 18. und 19. Jahrhundert', *Technologie und Politik*, vol. 16 (1980), pp. 155-179; B. Hoppe, 'Arbeit und Gesundheit. Historische Aspekte ihrer gegenseitigen Beziehungen', *Berichte zur Wissenschaftsgeschichte*, vol. 3 (1980), pp. 214-220.

32   Wolfram Kaiser and Werner Piechocki, 'Berufskrankheiten im Spiegel hallescher Universitätsschriften des 18.Jahrhunderts: die Bleiintoxikation im Hallescher Bergbau', *Wissenschaftliche Zeitschrift der Humboldt Universität Berlin*, mathematische-naturwissenschaftliche Reihe, vol. 17 (1968), pp. 767-772; Kaiser and K.Werner, 'Sozial- und arbeitsmedizinische Probleme in halleschen Inauguraldissertationen des 18. Jahrhunderts', *Zeitschrift für die gesamte Hygiene*, vol. 20 (1974), pp. 228-233.

33   Albert Kukowka, 'Prof. Dr. med. J.C.G. Ackermann, Arzt in Zeulenroda im 18. Jahrhundert — ein Pionier aus dem Gebiet der Berufs- und Gewerbekrankheiten', *Jahrbuch des Museums Hohenleuben-Reichsfels*, vol. 18 (1970), pp. 45-54.

34   Peter Schneck, 'Die gesundheitlichen Verhältnisse der Fabrikarbeiterinnen. Ausgewählte Aspekte der Situation in der sächsischen Oberlausitz im ausgehenden 19. Jahrhundert', *Jahrbuch für Wirtschaftsgeschichte*, vol. 3 (1975), pp. 53-72; Schneck, Die sozialmedizinischen Aspekte der Lage der Fabrikarbeiterrinnen in Deutschland im letzten Drittel des 19.

Jahrhunderts — dargestellt anhand der sozialen und gesundheitlichen der Verhältnisse der Fabrikarbeiterinnen in der sächsischen Oberlausitz, Diss. phil. (A), Dresden, 1975.

35   Dietrich Tutzke, 'Liebigs Beitrag zur Spiegelindustrie und der Kampf der Ärzte gegen die Hydrargyrose in den Spiegelfabriken', *NTM*, vol. 14 (1977), pp. 92-98.

36   Heinz Pilz, 'Kinderarbeit in der deutschen medizinischen Literatur zwischen 1800 und 1850', *Zeitschrift für die gesamte Hygiene*, vol. 21 (1975), pp. 694-699; Pilz, 'Kinder- und Frauenarbeit als arbeitsmedizinisches Problem 1842-1851', *24th International Congress of the History of Medicine, Budapest 1975, Acta* (Budapest, 1976), pp. 175-182; Pilz, 'Die Arbeitsmedizin auf den deutschen Hygienetagungen 1873-1883', *NTM*, vol. 15 (1978), pp. 45-55. For Moschke, see notes 16 and 44.

37   Karl-Heinz Karbe, Die Entwicklung der Arbeitsmedizin in Deutschland von 1780 bis 1850 im Spiegel der zeitgenössischen medizinischen Literatur, unpublished Habilitation thesis, Leipzig, 1978. Karbe, 'Zur Frühgeschichte des Kampfes gegen die Phosphornekrose in Deutschland', *Zeitschrift für die gesamte Hygiene*, vol. 22 (1976), pp. 447-454; Karbe, 'Die arbeitsmedizinische Forschung am Karl-Sudhoff-Institut als Beitrag zu einer marxistisch begründeten Sozialgeschichte der Medizin', *Wissenschaftliche Zeitschrift der Karl-Marx-Universität Leipzig*, gesellschafts- und sprachwissenschaftliche Reihe, vol. 29 (1980), pp. 564-574.

38   See the essays commemorating the 75th anniversary of the Karl-Sudhoff-Institut by Achim Thom, Irene Strube, Karl-Heinz Karbe, Klaus Gilardon and H.-H. Sauer in *Wissenschaftliche Zeitschrift der Karl-Marx-Universität Leipzig*, gesellschafts- und sprachwissenschaftliche Reihe, vol. 29 no. 6, (1980). A survey of recent GDR medical history is Thom, 'Zur Entwicklung der Medizingeschichte in Deutschland und in der Deutschen Demokratischen Republik', in Arina Voelker and Burchard Thaler (eds.), *Die Entwicklung des medizinhistorischen Unterrichts*, (Wissenschaftliche Beiträge der Martin-Luther-Universität Halle-Wittenberg (1982) no. 6), (Halle, 1982), pp. 24-49. For a schematic outline of FRG medical histories see Rolf Winau, *Deutsche Gesellschaft für Geschichte der Medizin, Naturwissenschaft und Technik. 1901-1976*, (Beiträge zur Geschichte der Wissenschaft und Technik, vol. 15) (Wiesbaden, 1978).

39   Walter Wuttke-Groneberg, *Medizin im Nationalsozialismus. Ein*

Arbeitsbuch (Wurmlingen, 1980), 2nd edn. (1982). Gerhard Baader and Ulrich Schultz (eds.), *Medizin im Nationalsozialismus. Tabuisierte Vergangenheit — ungebrochene Tradition?* (Berlin, 1980); Isa von Schaewen (ed.), *Medizin im Nationalsozialismus. Tagung vom 30. April bis 2 Mai 1982 in Bad Boll*, (Protokolldienst vol. 23/82) (Bad Boll, 1982); Projektgruppe 'Volk und Gesundheit' (ed.), *Volk und Gesundheit. Heilen und Vernichten im Nationalsozialismus* (Tübingen, 1982); Achim Thom and Horst Spaar (eds.), *Medizin im Faschismus. Symposium über das Schicksal der Medizin in der Zeit des Faschismus in Deutschland 1933 - 1945* (Berlin, 1983).

40   Sepp Gräsner, 'Neue soziale Kontrolltechniken durch Arbeits- und Leistungsmedizin', in Baader and Schultz, *Medizin im Nationalsozialismus*, pp. 145-151. Graesner, 'Leistungsmedizin im Nationalsozialismus', in Schaewen, *Medizin im Nationalsozialismus*, pp. 189-199.

41   Karl Heinz Roth, 'Pervitin und 'Leistungsgemeinschaft'. Pharmakologische Versuche zur Stimulierung der Arbeitsleistung unter dem Nationalsozialismus (1938 - 1945)', in Schaewen, *Medizin im Nationalsozialismus*, pp. 200-226.

42   Walter Wuttke-Groneberg, 'Leistung, Vernichtung, Verwertung. Überlegungen zur Struktur der nationalsozialistische Medizin', in *Volk und Gesundheit*, pp. 6-59; Wuttke-Groneberg, in Schaewen, *Medizin im Nationalsozialismus*, pp. 227-246.

43   Karl-Heinz Karbe, 'Das Betriebsarztsystem und zum Schicksal der Arbeitsmedizin im faschistischen Deutschland', in Thom and Spaar, *Medizin im Faschismus*, pp. 107-119. (also in *Zeitschrift für die gesamte Hygiene*, vol. 29 (1983), pp. 640-644.)

44   Gerd Moschke, 'Zur Behandlung der Berufskrankheitenfrage durch das faschistische Reichsarbeitsministerium bis 1936', in Thom and Spaar, *Medizin im Faschismus*, pp. 120-126.

45   Paul Weindling, 'Shattered Alternatives in Medicine' [Essay Review], *History Workshop*, no. 16 (1983), pp. 152-157.

46   The research project is called 'Dokumentation zur Geschichte der Arbeitsmedizin und des Betriebsgesundheitswesens in Deutschland ab 1835'. The project was carried out by Dietrich Milles, and directed by Rainer Müller and Alfons Labisch. Results are published in D.Milles and R.Müller (eds.), *Beiträge zur Geschichte der Arbeiterkrankheiten und der Arbeitsmedizin in Deutschland* (Dortmund, 1984).

47   See also Paul Klein et al., 'Zur Entwicklung der

Arbeitsmedizin in Deutschland bis zum Ende der Weimarer Republik', in Friedrich Hauss (ed.), *Arbeitsmedizin und präventive Gesundheitspolitik* (Frankfurt/M., 1982), pp. 28-40; Rainer Müller, 'Prävention von arbeitsbedingter Erkrankungen? Zur Medikalisierung und Funktionalisierung des Arbeitsschutzes', in Manfred Max Wambach (ed.), *Der Mensch als Risiko. Zur Logik von Prävention und Früherkennung* (Frankfurt/M., 1983), pp. 175-195. Müller, 'Grenzen und Reichweite der Arbeitsmedizin. Zu ihrer Geschichte, ihren Konzepten und Praktiken', in Horst Westmüller (ed.), *Gesundheitsrisiko Arbeitswelt. Aufgaben und Chancen einer arbeitsweltbezogenen Gesundheitsvorsorge,* (Loccumer Protocolle vol. 10) (Rehberg-Loccum, 1983), pp. 43-78.

48   Christian von Ferber, 'Gesundheitsvorsorge am Arbeitsplatz. Der Beitrag der Medizinsoziologie zu einem aktuellen sozialpolitischen Problem', *Die Betriebskrankenkasse,* vol. 68 (1980), pp. 225-230. On medicine and the civilizing process see: Alfons Labisch, ' 'Hygiene ist Moral — Moral ist Hygiene' — soziale Disziplinierung durch Ärzte und Medizin', in Christoph Sachsse and Florian Tennstedt (eds.), *Geschichte der sozialen Sicherung und der sozialen Disziplinierung in Deutschland* (Frankfurt/M., 1985); Labisch, 'Doctors, Workers and the Scientific Cosmology of the Industrial World: the Social Construction of 'Health' and the 'Homo Hygienicus' ', *Journal of Contemporary History,* vol. 20 (1985).

Part Two: Social Conditions and Risk Factors

3 FROM WORKERS' DISEASES TO OCCUPATIONAL DISEASES: THE IMPACT OF EXPERTS' CONCEPTS ON WORKERS' ATTITUDES

Dietrich Milles

The concept of occupational disease, prevalent in the Federal Republic of Germany, encompasses only a fraction of the extant dangers and injuries. Unless working conditions are taken into consideration the hazards can neither be alleviated nor recognised. This limited view of occupational health is a result of developments, in which 'experimental hygiene' and 'bacteriology' became predominant over social issues in public health. At the same time 'occupational diseases' have never lost their social stigma as diseases of disadvantaged groups. The assumption that occupational medicine was scientifically rather than medically based had a fundamental affect on concepts of occupational diseases. In place of 'workers' diseases', which bore a direct relationship to personal suffering and the social question, 'occupational diseases' became the focal point, which reflected a materialist and soulless working world. This meant that the science of occupational diseases was not value-neutral but subject to the needs of an industrial capitalist society. From this point onwards, the causal connection between labour and illness was in principle no longer questioned by medical science.

Criticisms of this modern view of occupational medicine persisted. The agitation of the workers' movements at the turn of the century resulted in a host of analyses of working conditions. The socio-political solutions proposed were all similar: the extension of social security, and improvements in safety at work and medical care. But these solutions were not related to the effect of the means of production on ill-health. The campaigners for social medicine who were closest to the workers' movement pressed for a widening of accident insurance and workers' compensation to include 'occupational diseases'. They significantly accelerated the reductionist trend of 'workers' diseases' being termed 'occupational diseases'. The 'occupational diseases' legislation of 1925 was, in this respect, a dead end in which individual improvements looked promising, but structural changes were more or less stillborn.[1]

In April 1980 the German Society for Occupational Medicine discussed the issue of 'work caused health impairment — fact or fiction?' This was not meant to be an exhaustive analysis of the entire subject, but rather a question of the radical critique of the crucial issue of 'work-caused diseases'. The Society intended to counter the radicals' challenge to the restrictive 'occupational diseases' model.[2] The conference heard more questions than answers in two papers concerned with the consequences of noise on health. The necessity for intensive study of the problem was made more than amply clear; but there was no coming to terms with the social dimension of the problem. The basic tenet of the conference was that the responsibilty of an accident insurance fund could only be established by proving a causal relationship between an illness and an occupation. The scientific method was dominant. Something other than a monocausal-scientific approach would 'dramatically change the system of social security'. The conference regarded occupational medicine's traditional principle of causality as having done a great deal to combat 'occupational diseases', because it necessitated scientific thought and argument.

The German Society for Occupational Medicine thus accepted a tradition which does not place the individual in a social context. The individual worker is severed from real social problems, and the treatment is laid down by guidelines determined by experimental and clinical medicine. At the same time, the Society omitted to mention that this tradition existed despite alternative medical opinions and socio-political interests.[3]

I

The medical profession of the nineteenth and twentieth centuries found it difficult to define the scope of medicine. The cause of an illness was not to be seen only in the breakdown of physiological processes or solved by clinical intervention. The social status of occupational medicine has meant that doctors are usually unwilling or unable to confront the capitalist-production methods which are damaging to health. The claim made by the medical reformer Rudolf Virchow during the revolutions of 1848 that 'the medical doctor is the natural attorney of the poor' has been overrated.[4] In the years of the 1848-49 Revolution doctors were extending their medical competence, and raising their income. The latter included payment for the medical supervision of the poor. The socio-political significance of these circumstances cannot be compared to the conditions of high industrialisation in the 1870s and 1880s.

The problem of relating illness to the system of production can be illustrated by the effects of mercury on the Fürth mirror workers. During the mid nineteenth century these workers were a frightening testimony to the health hazards of capitalist industrial production.[5] A series of shocking publications, which even the Reichstag could not ignore, followed a memorandum by the Fürth Provincial Medical Society.[6] This memorandum demonstrated that general practitioners were constantly confronted by the ill-health of these workers but they were unable to control the causes. The Fürth doctors maintained that the diagnosis of mercury poisoning was hampered by scientific and social factors. They revealed that the workers usually gave incorrect answers out of fear of their employer. Moreover, they accused many of their medical colleagues of a lack of concern because of financial interests. The report exposed the manufacturers' refusal to co-operate, suggested that the working conditions were better than they were, and showed that the owners behaved in a threatening fashion when they felt that they were being placed under scrutiny.

All those engaged in 'occupational hygiene' and social medicine have had similar experiences to those of the Fürth Medical Society. Of particular interest is the fact that the Fürth Medical Society recognised that the workers acted according to a logic of their own. The Fürth workers had become so accustomed to the mercury related illnesses that they merely treated themselves with gargles. Physicians saw these diseases only when the afflicted had to visit the doctor for some other reason. The workers knew very well that the only cure for the mercury tremor was to stop work, and that only then could medical attention help. The workers did not ignore medical treatment on principle, but came in droves, when a new method seemed to promise healing, as was the case with electrotherapy in the 1840s. No sooner did the treatment show little or no success, than they again kept away from the doctors' surgeries.

The physicians of the Fürth society found an unexpectedly high proportion of diseased people, and a frightening degree of risk: 'No one is in the mirror works for long without falling sick'. The doctors also learnt that the workers had a specific concept of their illness. Since the only cure was staying away from work, being ill was synonymous with an inability to work. Despite this, 'no mercury worker leaves his workplace prior to falling sick'. This means that the logic of the worker, expressed as a pair of scales, is calculated by him according to the degree of the weighted risks. While he can still work, the danger to his existence through loss of work and sustenance for him and his family is more serious than the 'not yet extant illness'. Only when the threat of loss of life predominates, does he acknowledge that he is really ill.

The spectacular study of the Fürth physicians was taken up by the Social Democrats to substantiate their criticisms of social conditions. The study also probably contributed to the abandonment of the mercury process a short time later. Despite the resistance of the manufacturers and entrepreneurs, a procedure was developed in which the poisonous mercury was substituted by silver, and one of the most horrific threats to life and health in the Fürth factories and elsewhere in Prussia was removed.

This example demonstrates a correlation between medical knowledge and socio-political developments: similar mechanisms operated in other less intensely threatened occupations. It was found that, as in Fürth, before health hazards directly affected workers they sought to adjust the labour process to working conditions so as to minimalise the danger. They also tried to heal themselves and each other, and to calculate the risks as well as the possibilities of change, as realistically as possible. The conceptualisation (thematisation) of work-caused illnesses can be assumed to be closely related to the chances of success of socio-political solutions. This is a rational perspective from the workers' point of view: the conceptualisation of dangerous diseases represents a coming to terms with an unavoidable threat to life. On the other hand, scientific knowledge and understanding of health risks, as in Fürth, could build on the situation of inadequate thematisation of the long-suffering workers. The ignorance of many general practitioners, who are confronted daily with the effects of industrial production is not explicable, except by a restrictedly positivistic view.

The socio-political consequences of recognition of the occupational hygiene problem was linked to the so called 'social question'. This 'social question' in the *Kaiserreich* referred to that which can be regarded as the costs of capitalist industrial development, and included the conflict with regard to pay, working hours, working conditions in factories, poverty, pauperisation, particularly in the cities, the question of lodgings for the workers, and the lack of social security for the old and infirm.[7] Added to this, came problems like cholera, public health and infant mortality. All these things had one thing in common, that those actually responsible for these ills, the owners and managers of the factories, simply shirked their responsibilities and passed them off onto society as a whole, while at the same time retaining their profits.[8]

In the political arena there was a parallel conflict between conservative and subversive forces. The supporters of the political workers' movement were identified as enemies of the state. There was a direct correlation between economic goals and conditions of the capitalist industrial system, and those of the Prussian-German monarchy, for which these very same goals and conditions were the basis of welfare and of national power.[9] At a time when the Social Democrats were characterised as traitors and assassins and were accordingly persecuted, the involvement of a physician with the 'social question' was deemed treacherous.[10] This was especially the case during the period of the 'anti-socialist laws' of 1878 to 1889, but continued only slightly abated until the dubious agreement of the Social Democrats to the war credits for the belligerent *Kaiserreich* on 4th August 1914. Even during the Weimar Republic, some sections of the bureaucracy continued to regard Social Democrats as revolutionary traitors to the fatherland.[11] One should not underestimate the pressure of public opinion. The overlapping of 'industrial hygiene' with political questions became acute towards the end of the nineteenth century. Theoretical developments in 'social hygiene', 'social pathology' and 'social medicine' bear witness to this.[12]

II

What degree of self awareness, and what form of understanding as to its scope, have developed in industrial hygiene? There are references to earlier views on industrial hygiene of individual physicians who based some of their findings on prior studies, thereby helping to provide some continuity. More concrete forms, such as specialist publications or institutes, were only attained by 'industrial hygiene' after the turn of the century. Yet, academic recognition was not forthcoming until after the Second World War. Industrial medicine still lacks identity as a scientific discipline and is consequently reluctant to delve into its own history or that of workers' diseases.[13]

If we take the Italian physician Bernadino Ramazzini (1633-1714) as the father of the infant 'industrial hygiene', then, the offence caused by his interest appears symbolic: he observed workers cleaning lavatories.[14] The preoccupation with questions of industrial hygiene has remained an offensive and objectionable matter within the medical profession up to the present day, and has generally just hovered on the fringes of science. Ramazzini was a general practitioner with an interest in effective help, and his methods of diagnosis involved direct investigation of workers. His work remained for over a century the most cited, much copied and many times translated basis for understanding illnesses caused by work. Concurrently with early nineteenth-century translations of Ramazzini, there were fears developing among sections of the bourgeoisie, as to the threatening social consequences of industrialisation.[15] Ackermann in 1780 already realised the necessity of focusing attention on the massing of young men in factories, since sickness could be generated and transmitted here more readily than elsewhere. He saw the reasons for this sickness not so much in poisonous or dangerous work-materials, but in poor nutrition.[16]

Developments in public and private medical care towards the end of the eighteenth century, were accompanied by a division of labour and industrial discipline that has had pernicious consequences to the present day. A concept of policing became dominant in the public sector while the domestic sphere became obsessed with 'hygiene', taking a moralistic approach which included a condemnation of those who could not afford to live a 'healthy' life.[17] This division into public and private medical care also resulted in a loss of awareness as to the political implications of 'industrial hygiene'. It was only with the issue of the fitness of military recruits that concern for the fitness of the workforce arose. The first laws for workers' protection forbidding excessive child labour and, later, female labour, grew out of this situation.[18] Working conditions in the factories, however, remained essentially a matter for the private sector, which meant that health hazards in the factories were also left to private initiative. The increase in industrial production in the middle of the century was accompanied by a rise in occupational diseases. These were registered statistically, and provided a basis for the renewed analysis of the relationship between work and illness.[19]

Halfort in 1845, borrowing not from Ramazzini but from Thackrah, systematically described the maladies of factory workers. He was aware of the fact that illnesses as well as the causes of illnesses multiplied with the expansion of the factory system.[20] Halfort maintained that the materials that were processed, and/or the method of processing, directly endangered health, and frequently led to death. He divided his study into occupational categories, to which he ascribed particular diseases, but he analysed the causal relationship in a general structural manner. Yet, it is just this analysis which runs counter to the capitalist mode of production. The interlocking of industrial production and health hazards was now the central issue.

Following further efforts, particularly by Friedrich Oesterlen, Ludwig Hirt recognised during the 1870s that there were limits to the usefulness of the concept of 'workers' diseases'.[21] If the analyses of social conditions producing disease were hindered by doctors remaining outside the factory, the resulting socio-political analyses were also weakened.[22] Hirt published a comprehensive account of 'workers' illnesses'. He recognised that although it received a polite reception, this did not imply a conversion to his views.[23]

At the beginning of the 1880s new developments occurred in response to changed social and political conditions. The Prague physician M. Popper recognised that there was a distinction between 'workers' diseases' and 'occupational diseases'. While the term 'occupational diseases' referred to the health hazards of specific occupations, the concept of 'workers' diseases' referred to those maladies which occurred more frequently in one class or stratum of society, irrespective of a specific categorisation or cause.[24] This differentiation on the one hand resulted in greater clarity, but on the other hand created the possibility with its emphasis on 'occupational diseases' that 'workers' diseases' could be relegated to the background. It is significant that such a displacement took place in connection with the development of social insurance.

III

The development of social insurance in Germany is cemented in those processes of socialisation and extension of law that Habermas termed 'inner colonisation'. The form of production including its harmful conditions and damaging consequences remained out of the reach of any controlling organisations or individuals, while the damaging consequences were suffered by society as a whole.[25]

The division between private ownership and public regulation for damages was the mainspring of the Liability and Social Insurance Laws in Germany. The Liability Law *(Haftpflichtgesetz)* passed the Reichstag shortly after the foundation of the Reich in 1871. The need for a legal remedy was acknowledged, but in practice it was so difficult to prove liability, that the claimant could only seldom derive any benefit. The effect of the law was that employers could not be made liable for risks to health in factories. The biting criticism of the Social Democrats demonstrated the inadequacy of legal redress, the uselessness of the Liability Law, which placed the burdens and risks of factory work on individual workers, and the refusal of society to take responsibility for damages, or to take preventive measures in the form of a state health service.[26]

These much praised laws achieved little more than a reform of the Poor Law on the basis of a semi-state social service. Accident insurance, which became part of the law after the Sickness Insurance Law of 1884, de facto transferred liability to medical insurance; there was a thirteen week interval before benefits were paid. These services were paid for by the worker: one part directly, and another indirectly, because the entrepreneurs treated it as wages in their book-keeping.[27]

The system of social insurance resulted in ever increasing legalism. As a result of the legal settlements, a network of dependence developed in a form analogous to Habermas' concept of client relationships. Individual claims of liability were categorised in a juridically structured procedure, which led to an ever increasing monetarisation and bureaucratisation of medical and nursing care.[28] The consequences of this system were disastrous. It had been intended to foster social integration, but it was too impersonal to do this. Sickness was synonymous with the inability to work and make a living, and carried the stigma of burdening others. Furthermore, the legal and financial conditions of sickness led to poverty, and in the long run it was seen as being anti-social. The monstrous consequences of the philosophy that every sick person is a burden on the healthy were seen in the actions of German Fascists.[29] Particularly in the case of accident insurance, we find that scientific debate about work-caused diseases has led to pressure from the grass roots for change. With only a few exceptions, the work of the doctor became predominantly that of providing advice and issuing certificates. Before the turn of the century this was equated with social medicine. Even today, those engaged in industrial medicine regard it as closely aligned to forensic medicine.[30] That such a tendency could, so easily, displace the socio-political role of social medicine was due largely to developments in 'hygiene' at the same time. We can only touch on the most significant developments.[31] Infectious illnesses and epidemics, rather than chronic diseases, were the problems which 'hygiene' had to solve. The pressing nature of this problem was demonstrated to the frightened populace by the cholera epidemics that persistently recurred until the 1890s. The solutions that 'hygiene' offered, and which functioned paradigmatically within the entire health care system, can be summarised as immunisation and sewers. Immunisation was the consequence of a bacteriological perspective, and was meant to protect people from external dangers to their health, while sewers were meant to carry away the filth that was the source of many dangers. In both cases there was a basic ignorance about the ecological and social conditions, and an unlimited trust in scientific and technical procedures. Today, hopefully, we see things differently. In the second half of of the nineteenth century there seemed to be no significant reason for thinking further about the health and socio-political implications of industrial mass production.

The interrelationships between the political, insurance, legal and medical-scientific developments of the 1880s led to liability determining cause and this became the focus of 'industrial hygiene'. An understanding of 'occupational diseases' pushed aside 'workers' diseases' and concentrated interest on the simple causal relationship between work and illness. The move towards the now dominant concept of 'occupational diseases', which already existed and was used extensively in the late nineteenth century prior to the Occupational Diseases' Statute *(Berufskrankheitenverordnung)* of 1925, is in principle a concession to the accident insurance. The rules for accident insurance stipulated that the accident insurance organisations were bound to pay damages in cases where the victim suffered either injuries not necessarily caused by the work environment, and resulting either in organic ailments or an impairment to his or her health. Damages were also due in the case of a sudden event. The transition of this 'accident' concept into accident insurance occurred without an exact definition on the part of the law makers. The concept initially excluded all work-caused illnesses which were the result of longer term factors.[32] The kernel of the 'accident' concept, the suddenness as well as the external causality are still used today and are regarded as valid.[33] Even today occupational physicians determine their areas of study according to the rules, regulations and premises of the Accident Insurance Law.[34]

We shall pursue the development from 'work-caused illnesses' to the concept of 'occupational diseases' a little further by looking at individual authors who are of significance to industrial hygiene.

IV

The two decades around the turn of the century saw a tremendous increase in industrial hygiene studies in Germany. This was the time of stocktaking, of producing monographs on individual branches of industry, institutional successes and handbooks.[35] This flowering of interest in industrial hygiene was no longer attributable to chance interest or personal idealism of a handful of individual researchers. Instead new impulses were given by the worker protection laws, and the new requirements of social insurance. These new stimuli were due to changed societal conditions, generated by mass public debates and an enlargement of the state social system. Following the repeal of the anti-socialist laws in 1890 efforts were made under the mantle of the social reformist 'New Course' to channel social conflict via reforms, such as the reform of industrial law and worker insurance.[36] The socio-political strategy enabled common bonds of analysis and strategy to be formed with 'industrial hygiene' research, which in turn could no longer be separated from political currents.

The Social Democrats had already begun, at the time of the anti-socialist laws to utilise reports of the factory inspectors (later to be called industrial inspectors — *Gewerbeaufsichtsbeamte*) to support their indictments and proposals in the Reichstag. 'Industrial hygiene' was thus brought into direct association with political questions involving worker protection. This became invaluable in the parliamentary debates about the the Protection at Work Bill of 1890-91. Industrial hygiene provided the central source for the numerous reports and studies about working conditions, such as the dangers to health inherent in individual branches of industry.[37]

At this juncture Ludwig Hirt's works on industrial hygiene warrant mention as he argued effectively for an improvement and up-dating of worker protection. Aiming at all-encompassing protection Hirt held that merely recognising that industrial labour led to illness was inadequate. Instead, he stressed the general and recurring factors responsible for ill-health. Hirt's questions dealt firstly with the materials to be processed, or those to be produced; secondly, with the posture and movements of workers; and thirdly, with the conditions and nature of the workplace. Through these issues he aimed at a more intensive study of occupational diseases, and more importantly at effective prevention. At the beginning of the 1890s Hirt published a popular text on 'worker protection' which summarised his earlier research.[38]

Theodor Sommerfeld, among others, followed this line of development in the interpretation of industrial hygiene. The sudden end of the 'New Course' and the socio-political reaction which set in in the mid 1890s put a stop to such positive developments and resulted in a renewed loss of political effectiveness. Further progress had to come from those quarters concerned with the development of social insurance and with the implementation of health and safety legislation, making these more significant than new laws.[39]

Again, the labour movement and social reformers provided the impetus. Socialists increased their involvement in the field of health insurance and thereby sought to gain a wider understanding of the particular dangers to which workers were exposed while at the same time acquiring medical facts for their political protests. The trade unions for their part built up worker councils *(Arbeitersekretariate)* to provide a form of legal aid and assistance during court proceedings with the social insurance bodies. Socialists collected industrial hygiene material of their own.[40] This strand in the development of industrial hygiene was limited by the insurance laws. The main area of contention was accident insurance, especially individual liability claims.

Within the medical profession, the few that concerned themselves with industrial hygiene reached the conclusion around 1900 that industrial activity could be dangerous to life and limb in two ways. Firstly, through an accident and secondly, via certain regularly recurring dangers closely connected to specific occupations. These distinctions were discussed mainly in relation to industrial poisoning since poisoning could be differentiated according to both categories. In the light of this, Rambousek already spoke of only occupational diseases when the malady stood in a close relationship to recurring harmful effects of a particular job.[41]

The essential criteria prescribed by the insurance laws already revealed the influence of industrial hygiene research. This is apparent in the question of timing: was it a matter of a sudden event or a continuing one over a long period of time? What was the causal connection between falling ill and one's job?

In their attempts to improve the situation of victims of occupational diseases and accidents, pragmatically thinking members of the medical profession sought to change the status of individual occupational diseases. The reformers wanted occupational diseases to be judged in the same manner, and by the same criteria, as was usual with 'accidents', as traditionally interpreted. New approaches were made in this direction by the Berlin toxicologist L. Lewin, who regarded the processes involved in occupational diseases, in particular occupational poisoning, as the sum of numerous smaller accidents.[42] The distinction between mechanical and toxicological occupational diseases was not settled legally in the way Lewin anticipated, although it later influenced the development of the concept of occupational diseases. An incentive to the analysing of diseases as accidents was that compensation for injuries was substantially higher than for sickness.

An increasing number of researchers recognised that the health of workers was threatened. From this stage onwards the two central questions that had arisen remained of primary importance. Theodor Weyl differentiated occupational hazards into accidents of an acute nature and into occupational diseases which were more or less chronic.[43] Coupled with this distinction was the demand that acute poisonings be recognised like accidents, and their victims compensated. Industrial hygiene research contributed a great deal towards such demands being acknowledged and taken into consideration in the discussion of codification and reform of social security. In the negotiations over the Reich Insurance Bill of 1911 the Social Democrat spokesman referred to industrial hygiene research when he demanded that the Accident Insurance Act be widened to include those illnesses which were caused or worsened by occupation. Yet, he mentioned only poisonings, and fell back on overseas experience. The Reichstag merely authorized the Bundesrat to extend accident insurance to include specific illnesses.[44]

In the Reichstag debates references to legal regulation in England over difficulties of determining the exact day on which an occupational illness occurred (as if an accident) played a decisive role. When the Second International Congress for Occupational Diseases met in Brussels in 1910 to prepare the equalisation of accidents and occupational illnesses before the law, English and German examples gave further guidance.[45] In 1891 an International Congress for Workers' Accidents had demanded that countries without invalid or old age insurance should be required to establish an insurance for serious accidents and occupational illnesses. Various international meetings heard the demands of industrial hygiene representatives, particularly Theodor Weyl and Ludwig Teleky, for equalisation of the two fields.[46] The Brussels Congress of 1910 posed the question of whether a separation of occupational illnesses and accidents at work was possible. Wilhelm Hanauer replied that the Accident Insurance Law erased any clear boundaries, since infectious diseases were regarded as accidents.[47] Furthermore the legitimate need for uniform regulation made it the duty of the law-makers to equate occupational diseases and accidents at work. For Hanauer this was overwhelmingly a legal and socio-political question.

As a consequence of these developments Theodor Sommerfeld and Richard Fischer were instructed by the International Association for Safety at Work to draw up an index of illnesses that warranted such equalisation.[48] Although the index was drawn up, further developments were hindered by the First World War. The Treaty of Versailles, however, pledged Germany to develop new laws that complied with the spirit of social and industrial justice.[49] This resulted in the Occupational Diseases Act of 1925 which included a list of eleven illnesses which were to be indemnified by accident insurance.

V

The interpretation of 'occupational disease' in the Reich Insurance Statute *(Reichsversicherungsordnung)* of 1911 was that of an illness due to harmful influences of work, which were not the result of an accident.[50] This legal construct influenced the concept of 'occupational disease' in such a manner that epidemiological arguments over specific health threats were excluded from the start. Emphasis lay on clinical proof. Compensation could be claimed only if an illness could be traced back to particular events at work.

Teleky concentrated on this legal handicap for reform of social insurance. He worked towards an effective practical method to decrease health risks and improve worker safety with the help of insurance.[51] He thought that social security organisations would protest if employers were made liable for general (and in origin not directly provable) damage to health. More effective, he thought, would be to introduce a sliding tariff for contributions of the employers to social insurance, according to the frequency of accidents occurring in their factories. Teleky believed that this, rather than general hygiene or technical requirements, would result in more direct improvements in worker safety. Compensation for provable occupational illnesses, as with accidents, would lead to new impulses and criteria for compensation and prevention. Furthermore, Teleky recognised that the injured worker was better off with accident insurance than with disability compensation. He demanded automatic liability from accident insurance as soon as an accident took place. Such a regulation, he maintained, could be achieved by the drawing up of a comprehensive register of occupational diseases. He had already realised in 1908 that such a register offered few problems for industrial hygiene and instead would enable immediate progress.[52] Teleky's goals were to improve the social position of the afflicted client groups, while at the same time providing an impulse for the prevention of industrial illnesses, and thereby upgrading the field of industrial hygiene. The realisation of these ideas in the form of the Occupational Diseases Act of 1925 nonetheless brought only marginal improvements for small numbers of those affected.[53]

In place of prevention came the medico-legal debate about causal liability resulting in industrial hygiene being pushed back to the fringes. Teleky realised in 1930 that the compromise reached between the employers and the trade unions was as good as worthless. Skin diseases, with the exception of a few well known occupational dermatoses, were ignored, and silicosis of the lung was restricted to those aspects which had been discussed in scientific journals.[54] It remained incomprehensible to physician and worker alike why the worker in a porcelain factory was compensated for silicosis but not workers in stoneware or earthenware factories. Teleky's strategy was to extend the registers for improved health protection. The political and legal ambiguity of the 'original causal-relationship' between work and sickness led in medical terms to unequivocal diagnosis. This absorbed the concept of the social responsibility of 'industrial hygiene' and to a large degree blocked the advance of preventive measures.

The history and origin of the development of the concept of occupational disease bears witness to the necessity of not applying to it what von Ferber terms the dubious dignity of perfection. We should rather look back at other effective alternatives that do credit to the social responsibility of 'industrial hygiene'.

Notes

1   L.Teleky, *History of Factory and Mine Hygiene* (New York, 1948); F.Koelsch, *Beiträge zur Geschichte der Arbeitsmedizin* (München, n.d.); Koelsch, 'Die Erforschung der Berufskrankheiten von der Wende des 18/19 Jahrhunderts bis zur Gegenwart', in E.W. Baader (ed.), *Handbuch der gesamten Arbeitsmedizin* (Berlin,

München, Wien, 1961), vol.2,1 pp. 37-68; H. Buess, 'Die Erforschung der Berufskrankheiten bis zum Beginn des industriellen Zeitalters', ibid., pp. 15-36; Buess, 'Über den Beitrag deutscher Ärzte zur Arbeitsmedizin des 19 Jahrhunderts' in W. Artelt and W. Ruegg (eds.), *Der Arzt und der Kranke in der Gesellschaft des 19 Jahrhunderts* (Stuttgart, 1967), pp. 166-178; K.-H. Karbe, 'Die Entwicklung der Arbeitsmedizin in Deutschland von 1780 bis 1850 im Spiegel der zeitgenössischen medizinischen Literatur', unpublished Habilitation Leipzig, 1978. On the history of hygiene see A. Fischer, *Geschichte des deutschen Gesundheitswesens* (2 vols., Berlin, 1933); T. Weyl, 'Zur Geschichte der sozialen Hygiene', in Weyl (ed.), *Handbuch der Hygiene,* suppl. vol. 4 (Jena, 1904), pp. 792-1062; A. Gottstein, 'Geschichte der Hygiene im 19. Jahrhunderts', in G. Stockhausen (ed.), *Das deutsche Jahrhundert* (Berlin, 1902), vol. 2, pp. 225-328; T. Sommerfeld, 'Die Gewerbehygiene', in S.N. Kreiss (ed.), *Fortschritte der Hygiene 1888-1913* (Berlin, 1913), pp. 134-208; F. Hueppe, 'Zur Geschichte der Sozialhygiene', in A. Gottstein, A. Schlossmann and L. Teleky (eds.), *Handbuch der sozialen Hygiene* (Berlin, 1925), vol. 1, pp. 1-70; M. Rubner, 'Die Geschichte der Hygiene', in M. Rubner (ed.), *Handbuch der Hygiene* (Leipzig, 1911), vol. 1, pp. 17-40; G. Rosen, *A History of Public Health* (New York, 1957); G. Rosen, *From Medical Police to Social Medicine: Essays on the History of Health Care* (New York, 1974); E. Lesky (ed.), *Sozialmedizin. Entwicklung und Selbstverständnis* (Darmstadt, 1977); H.-U. Deppe and M. Regus (eds.), *Seminar: Medizin, Gesellschaft, Geschichte. Beiträge zur Entwicklungsgeschichte der Medizinsoziologie* (Frankfurt/M., 1975); K.E. Rothschuh, *Konzepte der Medizin in Vergangenheit und Gegenwart* (Stuttgart, 1978); D. Milles and R. Müller (eds.), *Beiträge zur Geschichte der Arbeiterkrankheiten und der Arbeitsmedizin* (Düsseldorf, 1984).

2   W. Brenner, J. Rutenfranz. E. Baumgartner and M. Haider (eds.), *Arbeitsbedingte Gesundheitsschäden - Fiktion oder Wirklichkeit?* (Stuttgart, 1980); compare the report *Arbeitsmedizin und Ergonomie in der Bundesrepublik Deutschland* published by the Deutsche Forschungsgemeinschaft (Boppard, 1980).

3   For a critique see G. Elsner, F. Hauss, W. Karmaus and R. Müller, 'Arbeitsmedizin und Ergonomie in der BRD. Stellungnahme zu einer Denkschrift der Deutschen Forschungsgemeinschaft', *Jahrbuch für kritische Medizin,*

Argument-Sonderband, vol. 73 (1981), pp. 167-178; R. Müller, 'Zur Kritik der herkömmlichen Arbeitsmedizin', in M. Schmidt, R. Müller, F.-R. Volz, A. Funke and R. Reiser (eds.), *Arbeit und Gesundheitsgefährdung. Materialien zu Entstehung und Bewältigung arbeitsbedingter Erkrankungen* (Frankfurt/M., 1982), pp. 229-251; R. Müller, 'Prävention von arbeitsbedingten Erkrankungen? Die Medikalisierung und Funktionalisierung des Arbeitsschutzes', in M. Wambach (ed.), *Der Mensch als Risiko* (Frankfurt/M., 1983), pp. 176-195; F. Hauss (ed.), *Arbeitsmedizin und präventive Gesundheitspolitik* (Frankfurt/M., 1982).

4   On medical reform see K. Finkenrath, *Die Medizinalreform. Die Geschichte der ersten deutschen ärztlichen Standesbewegung von 1800-1850* (Leipzig, 1929); K.-H. Karbe, 'Über Forderungen deutscher Ärzte zur Verbesserung der Gesundheitsverhältnisse der Fabrikarbeiter im Jahrzehnt der bürgerlichen Revolution von 1848', *NTM*, vol. 8 (1971), pp. 45-53; E. Hansen, M. Heisig, S. Leibfried and F. Tennstedt et. al., *Seit über einem Jahrhundert...,Verschüttete Alternativen in der Sozialpolitik* (Köln, 1981).

5   A. Kussmaul, *Untersuchungen über den constitutionellen Mercurialismus* (Würzburg, 1861); F. Koelsch, 'Die soziale und hygienische Lage der Spiegelglas-Schleifer und -Polierer', in *Soziale Medizin und Hygiene*, vol.3 (1908), pp. 400, 483, 536; D. Hunter, *The Diseases of Occupations*, 6th edn (London, 1978), p. 311.

6   B. Schoenlank, 'Die Fürther Quecksilber-Spiegelbelegen und ihre Arbeiter', *Die Neue Zeit*, vol.5 (1887), pp. 145-167, 204-219, 256-266; B. Schoenlank, *Die Fürther Quecksilber-Spiegelbelegen und ihre Arbeiter. Wirtschaftsgeschichtliche Untersuchungen* (Stuttgart, 1888).

7   For the social history of Imperial Germany see H. -U. Wehler, *Das Deutsche Kaiserreich 1871-1918*, 2nd edn (Göttingen, 1975); W. Conze and U. Engelhardt (eds.), *Arbeiter im Industrialisierungsprozess. Herkunft, Lage und Verhalten* (Stuttgart, 1979); Conze and Engelhardt (eds.), *Arbeiterexistenz im 19. Jahrhundert. Lebensstandard und Lebensgestaltung deutscher Arbeiter und Handwerker* (Stuttgart, 1981); H. Pohl, *Forschungen zur Lage der Arbeiter im Industrialisierungsprozess* (Stuttgart, 1978); G. A. Ritter and J. Kocka (eds.), *Deutsche Sozialgeschichte. Dokumente und Skizzen*, vol. 2 *1871-1914*, 2nd edn (München, 1977); C.

Sachsse and F. Tennstedt, *Geschichte der Armenfürsorge in Deutschland. Vom Spätmittelalter bis zum 1. Weltkrieg* (Stuttgart, 1980); F. Tennstedt, *Vom Proleten zum Industriearbeiter. Arbeiterbewegung und Sozialpolitik in Deutschland 1800 bis 1914* (Köln, 1983); D. Langewiesche and K. Schönhoven (eds.), *Arbeiter in Deutschland. Studien zur Lebensweise der Arbeiterschaft im Zeitalter der Industrialisierung* (Paderborn, 1981).

8 Compare U. Mückenberger, *Arbeitsprozess—Vergesellschaftung— Sozialverfassung* (Bremen, 1981).

9 L. Machtan and D. Milles, *Die Klassensymbiose von Junkertum und Bourgeoisie. Zum Verhältnis von gesellschaftlicher und politischer Herrschaft in Preussen-Deutschland 1850-1878/79* (Frankfurt/M., Berlin, Wien, 1980).

10 H. v. Treitschke, 'Der Sozialismus und der Meuchelmord', *Preussische Jahrbücher,* vol. 41 (1878), pp. 637-647; P. Kampffmeyer, *Unter dem Sozialistengesetz* (Berlin, 1928); Kampffmeyer, *Zur Geschichte des Sozialistengesetzes. Artikel und Dokumente* (Berlin, 1928).

11 G. A. Ritter, *Staat, Arbeiterschaft und Arbeiterbewegung in Deutschland* (Berlin, Bonn, 1980); D. Groh, *Negative Integration und revolutionärer Attentismus. Die deutsche Sozialdemokratie am Vorabend des Ersten Weltkrieges,* 2nd edn (Frankfurt/M., Berlin, Wien, 1975); H. Mommsen, D. Petzina and B. Weisbrod (eds.), *Industrielles System und politische Entwicklung in der Weimarer Republik* (Düsseldorf, 1974).

12 L. Teleky, *Vorlesungen über soziale Medizin* (Jena, 1914); A. Grotjahn, *Soziale Pathologie* (Berlin, 1912); Grotjahn and I. Kaup (eds.), *Handwörterbuch der sozialen Hygiene* (2 vols., Leipzig, 1912-1913); A. Fischer, *Grundriss der sozialen Hygiene* (Berlin, 1913); R. Thissen, *Die Entwicklung der Terminologie auf dem Gebiet der Sozialhygiene und Sozialmedizin im deutschen Sprachgebiet bis 1930* (Köln, 1969); C. L. Trüb, 'Literarische Studie zur geschichtlichen Entwicklung der Begrifflichkeit (Definition) und Fachwörtlichkeit (Terminologie) Soziale Hygiene, Soziale Medizin und Medizin-Soziologie', *Öffentliches Gesundheitswesen,* vol. 39 (1977), pp. 233-246.

13 F. Koelsch, *Beiträge;* H. Schadewaldt, 'Arbeitsmedizin. Geschichte und Ausblick', *Medizinische Welt,* vol. 25 (1974), pp. 386-393; H. Jenss, *Zur Entwicklung des Jugendarbeitsschutzes von den Anfängen der Industrialisierung bis zur Gegenwart* (Frankfurt/M., 1980).

14   F. Koelsch (ed.), *Bernardo Ramazzini, der Vater der Gewerbehygiene* (Stuttgart, 1912); B. Ramazzini, *Untersuchung von den Kranckheiten der Handwerker und Künstler* (Leipzig, 1705).

15   J. H. G. Schlegel, *Die Krankheiten der Handwerker und Künstler, nach dem Lateinischen des Bernh. Ramazzini. Neubearb. v. Patissier. Aus dem Franz. übers. mit Vorrede u. Zusatzen* (Ilmenau, 1823).

16   J. C. G. Ackermann, *Bernhard Ramazzini's ... Abhandlung von den Krankheiten der Kunstler und Handwerker, neu bearb. u. vermehret.,...,* (2 vols., Stendal, 1780-1783).

17   C. W. Hufeland, *Gemeinnützige Aufsätze zur Beförderung der Gesundheit, des Wohlseyns und vernünftiger medizinischer Aufklärung* (Leipzig, 1794); Hufeland, *Die Kunst, das menschliche Leben zu verlängern* (Berlin, 1796); J. P. Frank, *System einer vollständigen medizinischen Polizei* (6 vols., Tübingen, 1778-1817); Frank, *Akademische Rede vom Volkselend als der Mutter der Krankheiten* (Pavia, 1790); A. Labisch, ' "Hygiene ist Moral — Moral ist Hygiene" — soziale Disziplinierung durch Ärzte und Medizin', in C. Sachsse and F. Tennstedt (eds.), *Geschichte der sozialen Sicherung und der sozialen Disziplinierung in Deutschland* (Frankfurt/M., 1985).

18   G. Anton, *Geschichte der preussischen Fabrikgesetzgebung bis zu ihrer Aufnahme durch die Reichsgewerbeordnung* (Leipzig, 1891; reprinted Berlin, 1953).

19   See Rainer Müller in this volume. D. Tutzke, 'Die Rolle der medizinischen Statistik in der Sozialhygiene in Deutschland vor 1933', *Zeitschrift für Ärztliche Fortbildung*, vol. 66 (1972), pp. 1158-1161.

20   A. G. L. Halfort, *Entstehung, Verlauf und Behandlung der Krankheiten der Künstler und Gewerbetreibenden* (Berlin, 1845); K. - H. Karbe, 'Zur Bedeutung von A. G. L. Halforts Werk über Entstehung, Verlauf und Behandlung der Krankheiten der Künstler und Gewerbetreibenden, Berlin 1845', *Zeitschrift für die gesamte Hygiene*, vol. 21 (1975), pp. 74-78.

21   F. Oesterlen, *Handbuch der Hygiene, der privaten und öffentlichen* (Tübingen, 1857); Oesterlen, *Handbuch der medicinischen Statistik* (Tübingen, 1865); H. Eulenberg, *Handbuch der medicinischen Statistik* (Tübingen, 1865); H. Eulenberg, *Handbuch der Gewerbehygiene. Auf experimenteller Grundlage* (Berlin, 1876); M. v. Pettenkofer and H. v. Ziemssen (eds.), *Handbuch der speciellen Pathologie und Therapie*, 3rd edn vol. 1 and *Handbuch der Hygiene und Gewerbekrankheiten*, 2 parts (Leipzig, 1882-1894).

22 L. Hirt, *Die Krankheiten der Arbeiter. Beiträge zur Förderung der öffentlichen Gesundheitspflege* (4 vols., Breslau, Leipzig, 1871-1878); Hirt, *Gesundheitslehre für die arbeitenden Klassen* (Berlin, 1891).

23 K. -H. Karbe, 'Ludwig Hirt - ein Kämpfer für den gesetzlichen Arbeitsschutz', *Zeitschrift für die gesamte Hygiene*, vol. 17 (1971), pp. 685-690.

24 M. Popper, *Lehrbuch der Arbeiterkrankheiten und Gewerbehygiene. Zwanzig Vorlesungen* (Stuttgart, 1882).

25 J. Habermas, *Theorie des kommunikativen Handelns*, (2 vols., Frankfurt/M., 1981); R. Voigt (ed.), *Verrechtlichung. Analysen zu Funktion und Wirkung von Parlamentarisierung, Bürokratisierung und Justizialisierung sozialer, politischer und ökonomischer Prozesse* (Königstein/Ts., 1980); W. Lepenies, *Das Ende der Naturgeschichte* (München, 1976); on social insurance see: F. Kleeis, *Die Geschichte der sozialen Versicherung in Deutschland* (Berlin, 1928); F. Tennstedt, *Soziale Selbstverwaltung. Geschichte der Selbstverwaltung in der Krankenversicherung*, vol. 2 (Bonn, n.d.); E. Wickenhagen, *Geschichte der gewerblichen Unfallversicherung*, (2 vols., München, Wien, 1980).

26 *Zeitschrift für Sozialreform*, no. 5-6 (1983), 'Hundert Jahre gesetzliche Krankenversicherung'; F. Tennstedt, 'Anfänge sozialpolitischer Intervention in Deutschland und England', *Zeitschrift für Sozialreform*, no. 9-10 (1983), pp. 631-648.

27 G. Schmoller, *Über einige Grundfragen der Sozialpolitik und der Volkswirtschaftslehre* (Leipzig, 1898).

28 J. Habermas, *Theorie*, p. 534.

29 T. Mason (ed.), *Arbeiterklasse und Volksgemeinschaft. Dokumente und Materialien zur deutschen Arbeiterpolitik 1936-1939* (Opladen, 1975); G. Baader and U. Schultz (eds.), *Medizin und Nationalsozialismus. Tabuisierte Vergangenheit-Ungebrochene Tradition?* (Berlin, 1980); T. Mason, *Sozialpolitik im Dritten Reich* (Opladen, 1977); A. Thom and H. Spaar (eds.), *Medizin im Faschismus* (Berlin, 1983).

30 K. Thomas, 'Wie es in Deutschland zum ersten Institut für Arbeitsphysiologie kam', *Arbeitsmedizin, Sozialmedizin, Arbeitshygiene*, vol. 9 (1967).

31 See note 1. K. D. Thomann, *Alfons Fischer (1873-1936) und die Badische Gesellschaft für soziale Hygiene* (Köln, 1980); H. Pilz, 'Die Arbeitsmedizin auf den deutschen Hygienetagungen 1873-1883', *NTM*, vol. 15 (1978), pp. 45-55;. C. v. Ferber, *Soziologie für Mediziner. Eine Einführung* (Berlin, Heidelberg, New York, 1975); W. Artelt and W. Ruegg (eds.), *Der Arzt und der Kranke in*

der Gesellschaft des 19. Jahrhunderts (Stuttgart, 1967); W. Artelt, E. Heischkel, G. Mann and W. Ruegg (eds.), *Städte-, Wohnungs- und Kleidungshygiene des 19. Jahrhunderts in Deutschland* (Stuttgart, 1969).

32   *Handbuch der Unfallversicherung in drei Bänden*, 3rd edn (Leipzig, 1909); *Dritte Verordnung über die Ausdehnung der Unfallversicherung auf Berufskrankheiten vom 16. Dez. 1936.Erläutert von M. Bauer u.a.* (Leipzig, 1937). H. Barta, *Kausalität und Sozialrecht. Entstehung und Funktion der sog. Theorie der wesentlichen Bedingung. (Empirische) Analyse der grundlegenden Judikatur des Reichsversicherungsamtes in Unfallversicherungssachen 1884-1914* (2 vols., Berlin, 1983).

33   F. Waterman, 'Berufskrankheiten und arbeitsbedingte Erkrankungen vor dem Hintergrund arbeitsmedizinischer Prävention der Berufsgenossenschaften', in W. Gitter, W. Thieme and H. F. Zacher (eds.), *Im Dienste des Sozialrechts. Festschrift für Georg Wannagat* (Köln, 1981), pp. 661-685.

34   G. Lehnert, Arbeitsmedizinisches Seminar (Köln, Lövenich 1979), p. 22 'Vom Befund zum Berufskrankheitenverfahren'.

35   O. Dammer (ed.), *Handbuch der Arbeiterwohlfahrt* (2 vols., Stuttgart, 1902-1903); E. Golebiewski (ed.), *Wegweiser der Gewerbehygiene (In Einzeldarstellungen)* (Berlin, 1898); A. Grotjahn and I. Kaup (eds.), *Handwörterbuch der Sozialen Hygiene* (2 vols., Leipzig, 1912); M. Mosse and G. Tugendreich (eds.), *Krankheit und soziale Lage* (München, 1913); E. Roth, *Kompendium der Gewerbekrankheiten und Einführung in die Gewerbehygiene* (Berlin, 1904); T. Sommerfeld, *Handbuch der Gewerbekrankheiten* (Berlin, 1898); T. Weyl (ed.), *Handbuch der Arbeiterkrankheiten* (Stuttgart, 1908); H. Albrecht (ed.), *Handbuch der Unfallverhütung* (Berlin, 1896); *Das Institut für Gewerbehygiene. Tätigkeitsberichte-Aufgaben* (Frankfurt/M., 1910).

36   D. Lindenlaub, *Richtungskämpfe im Verein für Sozialpolitik. Wissenschaft und Sozialpolitik im Kaiserreich, vornehmlich vom Beginn des 'Neuen Kurses' bis zum Ausbruch des Ersten Weltkrieges (1890-1914)* (Wiesbaden, 1967); U. Ratz, *Sozialreform und Arbeiterschaft. Die 'Gesellschaft für Soziale Reform' und die sozialdemokratische Arbeiterbewegung von der Jahrhundertwende bis zum Ausbruch des Ersten Weltkrieges* (Berlin, 1980); J. Reulecke, *Sozialer Frieden durch Soziale Reform. Der Centralverein für das Wohl der arbeitenden Klassen in der Frühindustrialisierung* (Wuppertal, 1982).

37   S. Poerschke, *Die Entwicklung der Gewerbeaufsicht in Deutschland*

(Jena, 1911); W. Bocks, *Die badische Fabrikinspektion.*
Arbeiterschutz. Arbeiterverhältnisse und Arbeiterbewegung in Baden
1879 bis 1914 (Freiburg, 1978).

38   L. Hirt, *Arbeiter-Schutz. Eine Anweisung für die Erkennung und*
Verhütung der Krankheiten der Arbeiter (Leipzig, 1879); Hirt,
Gesundheitslehre für die arbeitenden Klassen (Berlin, 1891).

39   T. Sommerfeld, 'Unsere sozialpolitische Gesetzgebung' in
Hygienisches Volksblatt (1902), p. 140. Sommerfeld, 'Die
Gesundheitsgefahr der Gewerbebetriebe', *Soziale Praxis*, vol. 13
no.37 (1904); Sommerfeld, 'Gewerbliche Erkrankungen der
Arbeiter', *Arbeiterschutz*, vol. 23 (1912), pp. 15; Sommerfeld, *Der*
Gesundheitsschutz im Betriebe (Berlin, 1922); Sommerfeld, *Der*
Gewerbearzt (Jena, 1905); Sommerfeld, *Atlas der gewerblichen*
Gesundheitspflege, (3 vols., Berlin, 1926-1928).

40   M. Martiny, 'Die politische Bedeutung der gewerkschaftlichen
Arbeiter-Sekretariate vor dem Ersten Weltkrieg', in H.O.
Vetter (ed.), *Vom Sozialistengesetz zur Mitbestimmung. Zum 100.*
Geburtstag von Hans Böckler (Köln, 1975), pp. 153-174; A. Müller,
'Arbeitersekretariate und Arbeiterversicherung in Deu-
tschland', Dissertation München 1904; R. Soudek, *Die deutschen*
Arbeitersekretariate (Leipzig, 1902).

41   J. Rambousek, *Gewerbliche Vergiftungen; deren Vorkommen,*
Erscheinungen, Behandlungen, Verhütung (Leipzig, 1911);
Rambousek, *Grundzüge der Gewerbehygiene und Unfallverhutung*
(Berlin, 1914).

42   L. Lewin, 'Die Vergiftungen in Betrieben und das
Unfallversicherungsgesetz', *Deutsche medizinische Wochenschrift*,
vol. 26 (1900), pp. 317-320. Lewin, *Die Grundlagen für die*
medizinische und rechtliche Beurteilung des Zustandekommens und des
Verlaufs von Vergiftungen und Infektionskrankheiten im Betriebe
(Berlin, 1907).

43   T. Weyl, *Handbuch der Arbeiterkrankheiten* (Stuttgart, 1908),
Einleitung.

44   F. Tennstedt, *Vom Proleten zum Industriearbeiter* (Köln, 1983), pp.
513ff; *Sitzung des Deutschen Reichstages. 16.5.1911, Stenographische*
Berichte über die Verhandlungen des Deutschen Reichstage, vol. 267
(Berlin, 1911), pp. 6802-6806.

45   W. Hanauer, *Gewerbekrankheiten und RVO* (Berlin, Lichterfelde,
1912). In general see W. Mommsen and W. Mock (eds.) *Die*
Entstehung des Wohlfahrtsstaates in Grossbritannien und Deutschland
1850-1950 (Stuttgart, 1982); G. A. Ritter, *Sozialversicherung in*

Deutschland und England. Entstehung und Grundzüge im Vergleich (München, 1983); N. Horn and J. Kocka (eds.) *Recht und Entwicklung der Grossunternehmen im 19. und frühen 20. Jahrhundert. Wirtschafts-, sozial- und rechtshistorische Untersuchungen zur Industrialisierung in Deutschland, Frankreich, England und den U.S.A.* (Göttingen, 1981); J. Alber, *Vom Armenhaus zum Wohlfahrtsstaat. Analysen zur Entwicklung der Sozialversicherung in Westeuropa* (Frankfurt/M., New York, 1982).

46 L. Teleky, 'Der I. Internationale Kongress für Gewerbe-krankheiten Bericht', *Zeitschrift für soziale Medizin*, vol.1 (1906), pp. 344-360; Teleky, 'Die Delegiertenversammlung der Internationalen Vereinigung für gesetzlichen Arbeiterschutz in Luzern, 28-30.9.1908', *Soziale Medizin und Hygiene*, vol.4 (1909), pp. 40-45; Teleky, 'Eine sozialmedizinische Kongressreise', *Wiener Klinische Wochenschrift*, vol. 24 (1911), pp. 21-66; Teleky, 'Gewerbekrankheiten und ihre Verhütung', *Vierteljahresschrift für Gesundheitspflege*, vol. 2 (1911), pp. 508-543.

47 W. Hanauer, 'Die Versicherung der Gewerbekrankheiten', *Zeitschrift für Versicherungsmedizin*, vol. 3 (1910), pp. 331-370; Hanauer, 'Ist eine Trennung der Gewerbekrankheiten von den gewerblichen Unfällen möglich, und welches sind die Unterscheidungsmerkmale?', *Aerztliche Sachverständigen-Zeitung*, vol. 20 (1910), pp. 405-429.

48 R. Fischer, *Der Entwurf einer Liste der gewerblichen Gifte. Bericht an die Int. Vereinigung f. gesetzl. Arbeiterschutz* (Frankfurt/M., 1910); T. Sommerfeld, *Entwurf einer Liste der gewerblichen Gifte* (Jena, 1908); *Internationales Arbeitsamt: Liste der gewerblichen Gifte und anderer gesundheitsschädlicher Stoffe, die in der Industrie Verwendung finden. Entworfen von T. Sommerfeld und R. Fischer* (Jena, 1912).

49 W. Schiff, *Der Arbeiterschutz der Welt. Eine Übersicht der Arbeiterschutz-Vorschriften aller Länder. Mit einem Anhang über die neueste Entwicklung des nationalen Arbeiterschutzes, über Arbeiterschutz im Friedensvertrag von Versailles und über die 1. Internationale Arbeitskonferenz in Washington* (Tübingen, 1920).

50 *Dritte Verordnung über die Ausdehnung der Unfallversicherung auf Berufskrankheiten vom 16. Dez. 1936. Erläutert von M. Bauer et al.* (Leipzig, 1937).

51 L. Teleky, 'Über die Grenzen der öffentlichen Gesundheitspflege in der heutigen Gesellschaftsordnung', *Die Neue Zeit*, vol.1 (1902-1903), pp. 178-185. Teleky, 'Die Krankenkassen und die Bekämpfung der Gewerbekrankheiten.

Referat auf dem 1. Internationalen Kongress für Gewerbekrankheiten in Mailand 1906', *Arbeiterschutz,* no. 16/17 (1907); Teleky, 'Gewerkschaft und Gewerbehygiene', *Der Kampf* (November 1907); Teleky, 'Die Gewerbekrankheiten und ihre Verhütung', *Ärztliche Zentralzeitung,* vol.23, no.14, (1911), pp. 111, 195, 223.

52 L. Teleky, 'Einige Bemerkungen zur Reform der Arbeiterversicherung', *Wiener Klinische Wochenschrift,* no. 47 (1908), p. 1635. Teleky, 'Der Gesetzentwurf über die Sozialversicherung vom Standpunkte sozialer Medizin', *Wiener Klinische Wochenschrift,* no.11-14 (1909), pp. 382, 419, 461, 497; Teleky, *Berufskrankheit und Unfall. Referat auf dem 2. Internationalen Kongress für Gewerbekrankheiten, Protokoll,* no.3-4(Brussels, 1910).

53 L. Teleky, 'Über Feststellung und Erkennung von Gewerbekrankheiten', *Klinische Wochenschrift,* no. 9 (1925), pp. 410-413; Teleky, 'Zur Versicherung der Berufskrankheiten', *Reichsarbeitsblatt,* no.14 (1925), p. 240; Teleky, 'Die Verordnung über Gleichstellung von Berufskrankheiten mit den Unfällen', *Klinische Wochenschrift,* no.37 (1925),pp. 1782-1785; Teleky, 'Arzt und Gewerbehygiene', *Ärztliches Vereinsblatt* (11.9.1925), p. 400; Teleky, 'Zur Durchführung der Verordnung über die Ausdehnung der Unfallversicherung auf bestimmte Berufskrankheiten', *Gewerkschaftszeitung,* no. 35-36 (1925), pp. 512-514, 524-526; Teleky, 'Die Versicherung der Berufskrankheiten', *Soziale Praxis,* vol. 41 (1927), pp. 1028-1032.

54 L. Teleky, 'Einige notwendige Reformen in der Versicherung der Berufskrankheiten', *Archiv für soziale Hygiene und Demographie,* vol. 5/6 (1930), pp. 442-451; Teleky, 'Zur Reform der gewerblichen Unfallversicherung', *Die Arbeit,* vol.11 (1931), pp. 859-868.

4 DISEASE, LABOUR MIGRATION AND TECHNOLOGICAL CHANGE: THE CASE OF THE CORNISH MINERS

Gill Burke

In this chapter I want to explore two of the questions that arise from historical examination of occupational diseases. Firstly, whether the introduction of preventive measures, or of compensation, should be seen as simply a response of sympathy or as the result of conflict between workers, employers and the state. Many examples suggest the latter is the case, even though the various parties may play differing roles over time. I will be discussing one particular occupational disease — phthisis/silicosis amongst metalliferous miners, taking the miners of nineteenth and early twentieth-century Cornwall as examples. I suggest that the incidence of lung disease amongst this occupational group was closely related to changes in the productive process and was not simply the inevitable consequence of hard rock mining. Also I stress that migration had a key relationship with disease amongst Cornish miners. I therefore examine the unique epidemic of ankylostomiasis that occurred there during the early years of this century. Secondly, and more briefly, I explore whether the process whereby any one disease becomes recognised as compensatory or indeed even as occupational has factors in common with the recognition process of other diseases? If so, can any general points be made from which perhaps a theory could be developed?

Mining in Cornwall — for tin and copper plus some lead and zinc, was a long established industry and the miners were men of acknowledged skill. Technologically the industry had been revolutionised by the steam pumping engines of Watt and Trevithick which enabled far deeper mining to be undertaken. Cornwall remained one of the world's major producers of copper and tin until the second part of the nineteenth century, when the crisis of severe price falls forced another revolution on the industry.[1] The demand for metals from the growing industrial economies of Europe and the United States had led to the development of new and greater sources of supply. The lifting of Tariffs on imported ores; the establishment of Free Trade; the opening of the Amsterdam Metal Market — all in the 1840s, signalled the expansion of world metalliferous mining on an unprecedented scale. Rapidly increasing supplies of copper from Chile led to a slump in price in 1866. This was followed by equally steep falls in the price of tin from 1873 as the easily mined deposits of Queensland, Australia, were exploited. For a hard rock, high cost producer such as Cornwall these price falls were catastrophic. Initially the effects of competition and price depression were countered in the traditional way — with closures of the more marginal mines and with large scale unemployment. However, as the price of tin continued to fall, new strategies were necessary if the industry were to survive. The first of these was technological innovation, then, when this proved insufficient, the industry itself was completely restructured. From 1895 most of the remaining mines had been formed into Limited Liability companies and were seeking outside investment. [2]

The technology turned to in the 1870s was rock drilling machinery. This, together with increased use of dynamite for blasting, following the expiry of Nobel's patent, meant that not only was there more dust in the mine air but also that the dust particles were very much finer. Since phthisis develops when dust containing free silica of less than 5 microns is regularly inhaled such changes were significant.[3] From 1875 the use of rock drills in the major mines enabled the Cornish companies to maintain output with a greatly diminished workforce.

The depressions of the 1860s and 1870s caused large scale migration from Cornwall. Indeed, one third of the population left the county between 1871 and 1881. Migration was no new phenomenon. From the late eighteenth century Cornish miners had been going and coming between Cornwall and other metal mining districts. Initially within England and Wales, then, as the mineral frontiers expanded, to other parts of the world. These migrations — to Central and South America from the 1820s; to the United States from the 1840s; to Australia from the 1850s — were mainly undertaken by men on their own. These single roving miners left wives and families behind in Cornwall. Their intention was to return. Sometimes they did so. Sometimes they sent for their families to join them overseas. Sometimes they moved on, from Mexico to California for example, when gold was found there in 1849. I have suggested that the single roving miner was a particularly Cornish phenomenon.[4] Family migration was far more likely to be depression pushed and permanent. Yet paradoxically the time of deepest depression in Cornish mining was the time that provided the last real opportunity for the single roving miner. The development of the South African gold fields from 1888 brought demand for skilled mine labour. By 1895 more Cornishmen were working in the Witwatersrand mines than were working in the mines of Cornwall.

During the last part of the nineteenth and the early part of the twentieth century, the links between Cornwall and South Africa were very close. Not only was 25% of the white workforce at the mines from Cornwall, but also there was considerable export of mining machinery and equipment. The Cornish firms of Harveys and Holman provided boilers, pumps and drilling machines for much of the initial expansion of the Rand, although later they were to be superseded. Harveys indeed maintained a Johannesburg office some years after the Anglo-Boer War. There were Cornishmen too in senior positions in the major mines. Men like Josiah Paull the manager of Ferriera Deeps, or R. Arthur Thomas who returned from the Rand to become manager of Dolcoath, the largest mine in Cornwall. In addition there were important geological similarities between the two countries. Cornish tin and Witwatersrand gold both lay in country rock containing a high degree of quartz and where lodes and ore contained very large concentrations of silica. The migration of the Cornish to the Rand and back thus set up a macabre pattern of the sub-migration of phthisis and its attendant disease of TB. This can be clearly seen in the *Health of Cornish Miners* Report 1904, where it was revealed that those who had worked in the gold fields of the Transvaal suffered from the highest death rate from phthisis of all miners investigated. Between 1900-1902 three hundred and forty two deaths were investigated in Redruth, Camborne, Illogan Gwennap and Phillack sub-regional districts of Cornwall (all mining areas). Of these, 185 had worked in South Africa and another foreign country.[5] This grisly migration had been recognised in Cornwall during the 1890s especially once men returned on the outbreak of the Anglo-Boer War. Many came home with their lungs ruined and died in such numbers that in Cornwall miners' phthisis increasingly became known as the 'Africa disease'.

The composite nature of phthisis is important not only in discussing the pathology of the disease, but also its incidence and migration between Cornwall and South Africa. It is also significant when considering what control measures were or were not imposed on mining operations in these two countries. Medical research into dust diseases of the lungs had declined during the later part of the nineteenth century in Britain. Experiments in the 1860s by E.H. Greenhow had confirmed that impregnation of pulmonary organs with dust was the result of the inhalation of vitiated air, but improvement in coal mine ventilation together with Koch's discovery of the tuberculosis bacillus focused medical attention on bacteriological causes away from non-medical prophylactic measures.

In South Africa it was not until the Transvaal Mines Regulation Commission of 1907-1910 that serious efforts were made to apply Greenhow's discoveries to underground conditions in the gold mines. [6] It then became increasingly obvious that the mixed dusts in the mine's atmosphere were what had a material effect upon the cause and development of phthisis. That such a complex composite disease was produced by variable factors suggests that its incidence clearly altered with the type of mining undertaken. There was strong evidence to show that both in Cornwall and South Africa the prevalence of the disease was inextricably linked to changes in the productive process. These in turn were related to the constraints imposed upon mining for profit under differing conditions of production. The enormous growth of the South African mines and their extensive degree of mechanisation initially masked the fact that similar innovations in Cornwall, albeit on a smaller scale, had the same deadly consequences. Miners' phthisis migrated with the miners between Cornwall and South Africa and back, but the disease could be contracted in either country.

This can be clearly seen from the number of deaths recorded in Cornwall during the earlier part of the 1890s whilst the great outflow of labour to South Africa was underway. Between three and four hundred men died of phthisis between 1893 and 1898. Similarly, the scale of the 'Africa disease' can be seen in the sharp increase in deaths following the outbreak of war. The 359 deaths from phthisis in 1899 rose to 490 in 1901.

Although the technological changes that led to greater dissemination of finer particles of dust placed all categories of underground workers at risk, it was the men who worked the rock drills who faced the greatest danger. To work a rock drill in Cornwall or South Africa during the 1890s was to face almost certain death. The only difference between the two places was the length of time involved. In 1902 it was calculated that a rock drill worker in Cornwall had an average of eight years work before dying of phthisis. In South Africa it was four years.

Of course, Cornwall and South Africa were not the only mining fields to be so affected. All over the world mechanisation had been introduced to metalliferous mines in varying degrees and with similar consequences by 1899. The rising incidence of phthisis deaths in the American and Australian mines led to labour agitation for better working conditions. These conflicts, coupled with expansion of the mining industries, led in time to provision of compensation, medical inspection, and the introduction of the axial feed drill — which made water an integral part of the drilling process. In South Africa these measures were also introduced following labour conflict, with however the added reaction of a greater use of black labour. These men too died in their thousands from phthisis. Only in Cornwall where the industry was rapidly declining were labour protests muted.

During this period of decline, the Cornish mines strove to retain their profitability by further cutting of working costs and neglect of working conditions. In the now transformed industry profitability was bought, more than ever before, at the expense of the health of the working miners. Furthermore, the Cornish mine companies resisted all attempts to get phthisis included in the schedules of compensatory illnesses. They argued that it could not be established that men had contracted the disease in Cornwall rather than South Africa. Even after 1918 when the Workmen's Compensation Act made silicosis a compensatable disease for miners, little attention appeared to be paid to the Act in Cornwall. However, there is evidence that between 1912 and 1916, miners from Cornwall, knowing they had phthisis, were going to South Africa to work in an attempt to get the lump sums offered in compensation at that time. Furthermore, for many years after 1918, the Miners' Hospital at Redruth in Cornwall was kept busy supplying x-ray evidence for compensation claims in South Africa.

The part played by the Cornish Miners, and the cost to them of sustaining the differing fortunes of the Cornish and South African mines had been recognised by Government researchers in 1909. A Report to the *Royal Commission on the Poor Laws* stated:

> Both in South Africa and Cornwall the nature of the
> work is conducive to phthisis in the case of the miner
> and to pauperism after his death of his widow and
> dependants. The conditions however vary in the two
> cases. In Cornish mines the wages are low and the
> conditions highly injurious. In the South African
> mines the conditions are very much more injurious
> but the rate of pay is high.[7]

The connection between Cornwall and South Africa and the migration of phthisis between the two mining areas enables comparisons to be made about the relationship between disease and the productive process, and about the extent to which ameliorative measures and compensation are introduced in particular circumstances. Unlike industrial/occupational accidents, diseases like phthisis have in the main a long time lag between exposure and symptom. This makes action by workers more difficult since at some distance in time such diseases may appear inevitable, not directly attributable to company policy, something that happens to individuals on a random risk basis. The relatively short time lag between rock drill work and death in Cornwall and South Africa at the turn of the century, the numbers of men involved and the dreadfully visible and protracted manner of their deaths left little doubt that this was more than simply an occupational hazard. Nonetheless, official response and action within the industry itself differed markedly between the two areas. Despite occasional downturns and the effect of the war, the South African mines were well able to respond to labour pressure and carry the costs of compensation and safety measures. A similar response in Australia at the mines of Ballarat and Bendigo underlined the importance of gold to the industrial world economy. The declining fortunes of Cornish tin mining however, generated a very different response. In 1899 the Inspector of Mines had explained his failure to insist upon the implementation of safety regulations by stating:

> I have not felt it expedient to press the matter as
> strongly as might be, for fear of arriving at the last
> straw which might bring about the total collapse of
> what little mining vitality remained.[8]

The industry resisted attempts by the state to enforce compensation in 1909 and ignored legislation after 1918. Although masks were made available for rock drill workers these proved difficult to work in and were unpopular with the miners. Yet, men who did not wear them, were subsequently 'blamed' for developing phthisis. Similarly, initial attempts to keep down dust through watering slowed down working speed — no small matter to men on contract — and forced the miner to work in mud and wet. Many men switched off the water. Phthisis then became their 'own fault'.

This tactic of 'blaming the victim' can also be observed by examining the ankylostomiasis epidemic in Cornwall. Similarly, there was differing reaction from workers and employers in areas of differing industrial fortunes and migration again played a key role. Ankylostomiasis is the disease resulting from infection by the hookworm *Ankylostoma duodenale*. The adult worms inhabit the duodenum and upper part of the small intestine and the enormous number of eggs produced by the female are passed out of the body via the faeces. If the faeces remain in a moderate temperature for a few days these eggs hatch and develop into larvae and worms. It is by swallowing the larvae, most usually by faeces polluted soil on the hands to the mouth, that humans are reinfected.[9] In tropical conditions where sanitation is poor, ankylostomiasis is still endemic. In the hot moist, muddy conditions of most Cornish and other mines (where sanitation was non-existent), the worm, once introduced thrived. The symptoms of ankylostomiasis are similar to those of anaemia—pallor, lassitude, dizziness, breathlessness, fatigue at smallest exertion. In addition infected persons might develop pustular eruptions on the skin. Although rarely fatal, the disease was not one that allowed sufferers to carry on hard manual work.

As had initially been the case with miners' phthisis, the pallor caused by ankylostomiasis was attributed to bad mine ventilation. 'Miners' anaemia' was thus seen as a worsening of the general debility resulting from mine work. The actual cause of the disease was first noted by Perroncito in 1882 as a result of post mortems carried out on men who died during the digging of the St. Gothard Tunnel in Switzerland. He also established that the disease was in existence amongst miners in Hungary, France, Belgium and Germany. So strong was the belief in the 'bad air' theory that few if any measures were taken to eradicate it. From 1900, the introduction of compulsory watering in the coal mines of Westphalia (to prevent coal dust explosions) provided an underground environment highly suitable for the spread of ankylostomiasis. Cases of the disease rose from 275 in 1900 to 1,355 in 1902.[10] The original infection was thought to have come with miners from Hungary, and a complete ban was put on foreign labour. The resulting labour shortage helped strengthen the demands made by the miners for the introduction of sanitary facilities and for payment whilst suspended from work due to infestation.

In Cornwall links had been made between migration and severe anaemia in the late 1850s and early 1860s before the existence of the parasite was known. A Penzance doctor had observed symptoms of severe anaemia amongst miners lately returned from Chile, and a St Just doctor made similar observations of other men. In 1898 the Chief Inspector of Mines drew attention to Perroncito's work in his Annual Report and alerted management and men to the true nature of the disease. Thus when men working in and around the New Engine Shaft at Dolcoath mine began to display symptoms of anaemia the manager, Mr R. Arthur Thomas, had little doubt as to the cause even though he took measures. The local Inspector of Mines still subscribed (as did many local doctors) to the 'bad air' theory, and it was on that ground that he applied to the Home Secretary for permission to have an enquiry conducted into the health of Cornish miners. The resulting investigation by J.S. Haldane not only revealed the incontrovertible relationship between rock drills and miners' phthisis, it also provided full details of the extent of ankylostomiasis in Cornish mines.

Almost certainly the worm had been brought to Cornwall by men returning from the tropics rather than from Europe. Although a few cases of ankylostomiasis had been noted at the Kimberley diamond mines, there were none in the very much drier mines of the Rand. A far more likely source of infection were the gold mines of Mysore, India. These had been expanded during the 1880s with the expiry of the old East India Company leasehold and the entry of British capital. One of the major mines on the Champion Lode was managed by the John Taylor Company, the world famous firm of mining engineers. This company had used Cornishmen in its undertaking since the first John Taylor had sent men to the Real del Monte mines in Mexico in 1823. There seems little evidence however of a complex relationship between Cornwall and India in the way there was with South Africa. The use of white miners was limited to shaft sinking and development work. The underground labour force at Mysore consisted of Indian men, women and children. By 1893 the period of expansion at Mysore was over. Indeed, the gold field was in eclipse beside the richer fields of the Rand. At this time many of the white miners were paid off, although some remained in a managerial capacity. Many of the men went from Mysore to the Rand, others came back to Cornwall.

It was at this time that cases of severe anaemia began to be admitted to the Miners' Hospital at Redruth. Of the 116 cases admitted between 1893 and 1902, well over half were from Dolcoath mine, and most of these were men who had worked in or around the New Engine shaft and sump. In particular the skin irritation and pus filled sores that preceded an attack became associated with sitting or leaning against rock or timber in that part of the mine and were called 'New Sump Botches'.[11] The Haldane enquiry revealed that nearly all the men working at Dolcoath and at many of the other Cornish mines, carried *Ankylostoma*. Many also carried other intestinal worms. These too were of tropical origins, and were similarly spread by ingestion of eggs or larvae. This was emphasised by the finding of the tape worm *Trichocephalus* rather than the much more common *Taenia solium*.[12] As *T. solium* is passed on via an intermediate host and therefore most often caught by eating infected meat, the presence of that worm would have related the disease to the miners' home and diet rather than their place of work and migration patterns.

The epidemic in Cornwall caused alarm, at a time when the similar but larger scale, epidemic in Germany resulted in labour unrest and loss of production. Fears were expressed that the disease might spread to the British collieries. Haldane was dispatched to Westphalia to seek further information on combative measures. His Report underlined what was already clear. That to prevent the spread of the disease it was necessary to have adequate underground sanitation. However, it was not the neglect of the mine companies in installing such facilities that was stressed. In almost all the many reports and articles on ankylostomiasis in mines, the blame was put on 'the filthy habits of the men'.[13]

> It must be clearly understood, however, that the prevention of the disease is largely in the hands of the men themselves and that it is their clear duty to observe and if necessary enforce, the precautions necessary for preventing the pollution of a mine.[14]

Fears that the disease would reach the coal mines proved groundless; only one case, in Glasgow, was reported amongst colliers. Yet the extent of infestation in the Cornish mines was considerable enough for several new cases a year to be reported for a decade after the Health of Cornish Miners Report of 1904.

This brief examination of occupational disease amongst metalliferous miners, suggests at the very least that the recognition of a disease as industrially generated and the introduction of compensation depends on a number of factors. These include the state of medical knowledge and the degree of effectiveness of organised labour, also perhaps the amount of time lag between exposure to risk and manifestation of the disease. Compensation for accident was altogether more straightforward, albeit that blame was more often laid upon workers for acting 'recklessly' than upon employers for their constant failure to provide a safe working *milieu* in which such recklessness would be unnecessary. For example, in case of accident arising from a miner riding to the surface in the kibble (ore haulage bucket) it was more likely that the miner himself would be prosecuted for breaches of the 'safety regulations' than the mine company for not providing any means to mechanically raise the men to surface. Cornish miners continued to face a thousand feet of ladder climb after their day's work, long after lifts had been installed in the collieries. None the less, the result of accident was immediate and obvious, the link between changes at work and increases in deaths from disease was often more difficult to perceive. The two most important factors in the recognition of a disease as both industrially generated and compensatory would appear to be, firstly, the importance of an industry to the national economy, and, secondly, that industry's current condition. The differing response to phthisis compensation in Cornwall and South Africa exemplifies this. These factors provide the context in which labour agitation for compensation may or may not be successful. Similarly, fears of labour unrest in vital industries may generate action by the state — as in the case of the ankylostomiasis epidemic in Cornwall where fears of the disease spreading to the far more crucial coal industry provoked government enquiry and action. Given the Westphalian example, the government can have been in no doubt that had the disease spread, there would have been considerable unrest with expense to the industry, and possibly the state, arising from loss of earning payments. Although treatment for ankylostomiasis was compulsory in Cornwall, no financial compensation was paid there. The state of the industry together with the established pattern of migration precluded effective labour militancy. It was not until the outbreak of war in 1914 closed the migration escape route that trades unionism became established in the Cornish mining industry and the following years were marked by industrial conflict, but even then the decline of the industry rendered such conflict ineffective at other than purely local level.

It is possible to suggest that a typology could be drawn up mapping out the career of a disease from acknowledgement of its industrial causes through to the establishment of state compensation schemes and the subsequent working of such schemes in practice. Certainly a brief overview of historical and contemporary studies of miners and quarrymen here and elsewhere suggests such a typology for mining dust diseases. To establish such a typology would not only increase our historical understanding, but would also enable wider questions of policy to be acknowledged and addressed.

Notes

1   G.M. Burke, 'The Cornish Miner and the Cornish Mining Industry 1870-1921', University of London PhD 1982.
2   G. Burke and P. Richardson, 'The Decline and Fall of the Cost Book System in the Cornish Mining Industry 1895-1914', *Business History*, vol. 23 no. 1 (1981), p. 4.

3   W.R. Parkes, *Occupational Lung Diseases* (London, 1974), pp. 166-216.

4   G. Burke, 'The Cornish Diaspora in the Nineteenth Century' in S. Marks and P. Richardson (eds.), *International Labour Migration: Historical Perspectives* (London, 1984), pp. 57-75.

5   PP 1904 XIII (Cd 2091) *Report on the Health of Cornish Miners.*

6   G. Burke and P. Richardson, 'The Profits of Death: a Comparative Study of Miners' Phthisis in Cornwall and the Transvaal 1876-1918', *Journal of Southern African Studies*, vol. 4, no. 2 (1978), p. 149.

7   PP 1909 *Royal Commission on the Poor Laws* xvi (Cd 4653) Appendix IX. Final Report on the relation of industrial and sanitary conditions to Pauperism by A. D. Steel Maitland and Miss Rose E. Squire.

8   PP 1899 5,xv (c 9264-vi) Report of HM Inspector of Mines.

9   A.O. Lucas, *A Short Text Book on Preventive Medicine in the Tropics* (London, 1976), pp. 132-137.

10   PP 1903 XV (Cd 1671) *The Epidemic of Ankylostomiasis in the Westphalian Colliery District* by Mr T.R. Mulvany, HM consul General at Düsseldorf and Dr F. Ph. Koenig British Vice Consul.

11   A.E. Boycott and J.S. Haldane, 'An Outbreak of Ankylostomiasis in England. No 1', *Journal of Hygiene*, vol. 3 (1903), pp. 95-136, 104.

12   A.E. Boycott, 'Further Observations on the Diagnosis of Ankylostoma Infection with Special Reference to Examinations of the Blood', *Journal of Hygiene*, vol 4 (1904), pp. 477-479.

13   Ibid. p. 479.

14   PP 1902 XVII (Cd 1318) *Report on an Outbreak of Ankylostomiasis in a Cornish Mine* by J. S. Haldane.

5   T.N.T. POISONING AND THE EMPLOYMENT OF WOMEN WORKERS IN THE FIRST WORLD WAR

Antonia Ineson and Deborah Thom

At the end of the First World War, a new role was proposed for medicine, a role which arose from the experience of dealing with poisoning and toxic jaundice among shell filling factory workers using tri-nitro-toluene, TNT. An article in the *British Medical Journal* described the war-time history of jaundice as 'an object lesson showing the grounds on which scientific medicine should be based in the future'. Observation and experiment, in the laboratory using animals and in medical practice using people, was to be the foundation of the new medicine.

> The medicine of the future will attain that perfect advancement and full knowledge which all desire by the association of the physician and the scientific worker, not only in the laboratory, but also at the bedside.[1]

The Health of Munition Workers Committee and medical staff at the Ministry of Munitions added the factory to the list. The new practice would encompass an alliance between doctors and factory managers grounded in the experience of the war;

> The success of the special work of the factory medical service, together with the help given by research during the war ... opened up a new field for co-operation between medical science and the factory management of the future.[2]

In this paper we are to examine the response to TNT poisoning among filling factory workers. About 50,000 workers were employed on filling at a time, and 100,000 during the course of the war. The vast majority of them were women. We shall argue that the role of medicine was very far from the image of scientific advance outlined above. TNT poisoning, in common with other aspects of munitions production during the war, was the site for a struggle for control. What was unusual was the extent to which medical and managerial interests became clearly combined.

The responses to poisoning can be divided into two groups. One, originating with those organising the production of filled shells, included medical, technological and managerial responses. The other comprised trade union action and individual or group responses among those working with TNT. The notion of what TNT poisoning consisted of was articulated through a wide variety of experiences, those of the people who became ill on the one hand, and of the efforts of factory managers and the Ministry of Munitions to organise and control the filling of shells on the other. The medical view of TNT poisoning cannot be separated from the incorporation of factory doctors into the management of filling factories or from a laboratory research programme with the over-riding aim of efficient shell production.

Shell filling using TNT was established on a large scale for the first time in the war. Previously, almost all filling had been done at Woolwich Arsenal; a small amount for export was done elsewhere. TNT was being introduced to replace other explosives in the pre-war years, partly because it was thought to be less toxic than dinitrobenzene and was less explosive than lyddite. At the beginning of the war filling was carried out almost entirely manually. The work was physically hard, repetitive, and there was a constant danger and fear of explosion. The workers were subject to strict rules of behaviour, in addition to the new systems of management being introduced in munition work in the war. The Ministry of Munitions described the work as particularly suitable for women, as they were not seen as minding its unskilled, monotonous and dead-end nature — it suited their temperament.[3] The Ministry soon recognised that shell filling was far more dangerous to the workers' health than any other munitions employment, but there was an almost total lack of attention paid to the coincidence of a concentration of women workers with a singularly dangerous task.

The connection between TNT and deaths from toxic jaundice reported among filling factory workers in 1915 was quickly made, largely because of similarities with the action of other industrial poisons. Dinitrobenzene, used in the dyeing and explosives industries, had been known to be linked with toxic jaundice for some years, and tetrachlorethane, a constituent of aeroplane dope, was similarly linked with jaundice soon after the beginning of the war. Toxic jaundice was made a notifiable disease as a result of the dope case in January 1916. By this time, public knowledge of the effects of working with TNT was proving a problem for the Ministry. Workers were refusing to take jobs in filling factories, those already employed becoming 'disorganised through fear of contact' and levels of absence from work on grounds of sickness were said to be high.[4] Some action had to be taken if the production of shells was to be kept up.

The result was a combination of managerial and medical solutions to the problem. The Health of Munition Workers Committee, which covered all munition work, issued a memorandum entitled *Special Industrial Diseases* in February 1916, including a section on TNT poisoning. The Ministry of Munitions produced regulations in September of the same year. Sir George Newman, as Chairman of the Health of Munition Workers Committee, inspected women TNT workers at Woolwich Arsenal in July 1916, and found a high level of illness. 37% of the women experienced what he described as 'severe pains below the xiphisternum, associated with loss of appetite, nausea and constipation', and 25% had dermatitis. 36% suffered from depression, 8% from irritability, and 34% experienced some change in menstruation. [5] In the following month, Christopher Addison, then Minister of Munitions, called a meeting of representatives of all departments of the Ministry concerned with TNT, and a TNT Advisory Committee was appointed in October. Members included people from the Medical Research Committee, various sections of the Ministry, the Factory Department of the Home Office, and the Health of Munition Workers Committee. The Advisory Committee produced a new set of regulations in February 1917, and issued recommendations on the duties of factory medical officers, the use of respirators and so on. These were largely based on research set up by the Medical Research Committee (MRC), which had begun a series of experiments of the absorption of TNT in August 1915. The Advisory Committee discussed reducing the contact between workers and the poison by alternating employment on TNT with so-called clean work, mechanisation of filling, the use of respirators and other protective clothing, and exhaust ventilation of the atmosphere in factories.[6]

It is extremely difficult to estimate the extent to which the recommendations and regulations on work with TNT were actually carried out, or the degree to which they were responsible for the reduction in deaths from TNT poisoning due to toxic jaundice in 1917 and 1918. Certainly the number of deaths did fall — from 52 in 1916 to 44 in 1917 and 10 in 1918 — but it is arguable that this was not the result of the medical attempt to investigate and combat the disease.[7]

From 1916 information about the effects of working with TNT were censored in both public newspapers and in the medical press. The results of inquests could only be published in a brief, standard form so that recruitment of labour was not hampered.[8] The reportings of medical research also had to pass the Press Bureau censorship, and this limitation seems to have passed without objection from the medical profession. The aim was clearly to maintain the state of ignorance in which, according to the Chief Medical Officer of the Ministry, munition workers began TNT work.[9]

Factory doctors and those involved in managing the munitions factories accepted that the prime necessity was to produce shells as efficiently as possible. The effects of this on the nature and direction of the medical investigation of TNT poisoning were clear. Doctors concentrated on distinguishing between symptoms of poisoning which did or did not develop into toxic jaundice and possible death. It was assumed that work with TNT was likely to lead to illness of some degree; the point was to keep those workers whose lives were not endangered filling shells and to remove the early cases of toxic jaundice. However this was by no means a straightforward task. The list of early effects of work on TNT was long; it included drowsiness, frontal headache, eczema, dermatitis, loss of appetite, gastritis, constipation, cyanosis, shortness of breath, vomiting, anaemia, palpitation, yellow or orange staining of the skin and hair, depression and a metallic taste in the mouth. Attempts were made to divide this list into categories, for instance two medical officers working among women filling factory workers suggested in an August 1916 issue of the *Lancet* that there was one group of irritative symptoms, which may lead, in time, to toxic symptoms.[10] Some of the effects were said to be positively or negatively correlated with the susceptability of the workers to toxic jaundice. Dermatitis was said to be inversely related to poisoning of other kinds, due to variations in the ease of passage of TNT through the skin and into the blood, for instance.[11]

The relationship between 'serious' and 'minor' effects remained unclear throughout the war; it was claimed by the Medical Research Committee that the reduction of deaths was evidence of the reduction in illness, and various ratios of one to the other were proposed.[12] Once the number of deaths had begun to fall, increasing attention was paid to the economic effects of 'minor' illness.

Linked with this attempt to distinguish between serious and minor symptoms was the belief that certain people 'have an idiosyncracy towards toxic absorption.'[13] This was the basis for the official guidelines issued by the Ministry of Munitions on TNT work, and many proposals were made by doctors as to the characteristics of the so-called 'susceptibles'. The latter included the young and the old, those who had gastric or liver illness, alcoholics, people with syphilis, those who sweated a lot, the malnourished and the over-fatigued — a list which must have included most of the filling factory workers. The ideal aim was to be able to identify and exclude 'susceptibles' at a pre-employment screening by the factory doctor, at the same time as establishing that the worker was in an adequately healthy state for the job. However no reliable means of identifying 'susceptibles' was found, although eye colour and general health were used. The other function of this inspection was to identify existing illness, not necessarily or even usually for treatment or to exclude the person from work, but often so that the illness could not be used to claim compensation as being caused by TNT.[14] The factory doctor was supposed to have no curative role.

An MRC report published in 1921 concluded that

> it has been suggested at different times that alcohol, syphilis, adenoids, obesity and bad feeding are predisposing causes, but no evidence is available pointing in any of these directions.[15]

The Health of Munition Workers Committee agreed but continued to support the importance of *some* individual characteristic being responsible in the *Final Report* ;

> the few affected are not always those who, owing to ill-health or malnutrition might be expected to be especially liable. Industrial conditions, though important, have perhaps less influence than personal idiosyncracy. [16]

Once workers had been accepted for munition work, factory doctors, appointed by the Ministry of Munitions, were supposed to carry out regular checks on TNT workers and to withdraw those suffering from poisoning from work. The problem was to differentiate between those who were in danger of becoming 'seriously' ill with toxic jaundice, and those who were merely suffering from 'minor' effects. In September 1916 a TNT 'facies' was described, as typical of those who should be removed from work:

> a pale face lacking in expression and like anaemia but peculiar in itself, lips that can hardly be described as cyanosed but of an ashen blue colour, similar gums, and perhaps a faint trace of yellow on the conjunctivae, the rest of the skin showing no icterus.[17]

The same writer later stressed that it was important that the doctor should observe people at work;

> The excitement of going to see the doctor — over
> which at present a great deal of munition time is
> wasted — tones up the patient and disguises the
> symptoms. The best way is to steal round the
> workshops ...[18]

The worker's own experience of ill-health had little or no place in
the diagnosis of early poisoning. No questions were to be asked. An
article in the *Lancet* in 1916 claimed that even among those already
affected, 'the history given by a patient is often very misleading. Many
of the workers have no idea as to the nature of the substance upon
which they are working . . .'[19] Medical examinations wasted time
which could be devoted to shell filling, some argued, but others stressed
their importance in reassuring workers that care was being taken of
them.

Allied to the problem of the definition of the seriously ill was the
exclusion of some effects of TNT from medical consideration at all.
The most obvious example was the yellow staining of skin and hair,
which was of concern to the people affected but to doctors was simply a
sign that the person worked with TNT. The effect on menstruation,
mentioned by Newman early in the war, was not studied in any of the
later research. Although it would be impossible (and is probably
undesirable) to attempt retrospective diagnosis, it seems quite clear
from accounts by women workers that they experienced much more
general slight illness than could be suggested by the level of notification
of toxic jaundice.

Proposed treatments of TNT poisoning if jaundice was not present
consisted of some variant of bed rest, a milk diet, and keeping the
bowels open. Jaundiced patients were to be given alkali-producing
drugs, linseed and mustard poultices for the liver, and rectal and
intravenous saline injections. It was generally admitted, however, that
the prognosis was poor and that the treatment was not based on any
great understanding of the condition. The medical workers said that
even if the person survived, their health was likely to be permanently
damaged.[20]

The factory medical officers were not only responsible for
inspections of workers before and during employment but were also
able to advise managers on working techniques from the point of view
of health. They were supposed to be seen by workers as being
responsible for protecting them against poisoning. The importance of
limiting doctors' activities and of ensuring their allegiance to
management was recognised by the TNT Advisory Committee at its

first meeting, in 1916; 'Doctors should be paid by the Factory Managements, otherwise the highest factor in the authority of the Management would be removed'. The doctor had to balance losing 'a few lives in the manufacture of TNT' with the importance of maintaining the supply of shells.

> If 10 percent of the workers at a factory were knocked off because they were susceptible, there would be such a panic that the Factory would probably lose its labour. Of 200 people recently engaged for Perivale Factory, only 21 came in when they found it was a filling factory. Leeds factory was already losing its labour at the rate of 200 a week. . . . [Doctors] should work hand in glove with the management, and should not pull a single girl out, except with the consent and approval of the Factory Management. Panic should be stopped by convincing the operatives that the Ministry had 'got the thing under control'. [21]

The doctors' function was to remove workers who were particularly likely to die, as long as there were not too many of them, and to deal with deaths, post-mortems and compensation. In the latter case the point was to reduce payment for 'unnecessary' compensation claims by excluding other causes of death at post-mortem, and to give evidence at the inquest 'as to the precautions taken in the factory to protect the workers'.[22] There are few records of the details of compensation cases, but in one example which does exist the doctor changed his diagnosis from TNT poisoning to pneumonia once the man had died; a letter to the doctor from the staff superintendent at Chilwell, where the man had worked, suggested that,

> perhaps your certificate (of TNT poisoning) was only tentative. You will readily appreciate of course that we imply no criticism whatever of your care of the patient, but this question of TNT poisoning is one of national importance and that is why we labour this point of diagnosis.[23]

Factory medical officers, therefore, acted as medical administrators governing the passage of workers in and out of contact with TNT, according to certain preconceptions about the nature of the action of TNT and about the relative importance of the health of workers and the production of shells. Their work was supported and informed by more basic research, mainly carried out from August 1915 by a group headed by Dr Benjamin Moore of the Applied Physiology section of the Medical Research Committee. This work was intended to elucidate the route of entry of TNT to the body, and from the start was closely linked with decisions about the use of protective clothing (such as gloves, aprons and boots), respirators, and exhaust ventilation.

Tests for the detection of TNT poisoning were developed, and systems of alternation of employment proposed. This example of the co-ordination of laboratory research with clinical medicine and factory management was coloured by the stress on maintaining output, if necessary, at the cost of the health of the individual. Moore was a major supporter of the theory that individual idiosyncracy lay behind TNT poisoning, and claimed that if enough workers were screened, exposed to TNT and selected according to their reaction, a naturally resistant workforce would result through the 'industrial selection of the fittest'.[24]

The connection between the scientific research and the development of new techniques of working was close, although the major change from hand-filling to machine filling of shells took place under the impetus of increased production rather than reduced contact between worker and TNT. Medical support for the use of respirators, for example, was privately admitted to rest on dubious evidence but was publicly stressed as part of the workers' own responsibility for protecting themselves from poisoning.

A similar alliance of doctors and management lay behind organisational changes in shell filling; alternation of employment on and off TNT work was partly based on Moore's work and could be controlled by factory doctors. It could also be controlled, unofficially, by women themselves, who were for once in a situation where their labour was in demand. Their attitude to their work in general, and their willingness to keep working on filling affected their control over poisoning more than their beliefs about the dangers of TNT.

Did munition workers in fact ignore TNT poisoning as an issue? They experienced it in the context of the first mass mobilisation for warfare in Britain which included women, and their involvement in war work was characterised as ancillary to male fighters. The slogans of recruitment posters added to 'Do Your Bit', the slogan 'Replace a Man for the Front'.[25] Male munition workers were released for combatant service as soon as they could be replaced by women or boys. Munitions work was perhaps the war work par excellence and this has confused the records of the experience — so much so that there are no adequate contemporary figures for the relative distribution of substituted and diluted workers since the interests involved in the compilation of such statistics were heavily loaded in defining the nature of the work or the definition was subject to much negotiation. (A substitute replaces a man directly, doing the same work. A dilute replaces a skilled man and does the same job but often has the work re-arranged or uses adapted machinery). It is clear though that the majority of women on TNT were working in jobs newly created for war and only extant for the duration; the majority were employed in government filling factories (only operational 1916-1918) and, therefore, on new jobs with largely female workforces, in areas of the country previously unindustrialised. Most TNT workers were apparently new to industry, the majority from domestic service or agriculture (unlike women engineering workers).[26] They entered a new style of production that had been designed to accommodate inexperienced workers. Reorganisation of the work process involved increasing repetition. Payment systems were devised to keep output high and rising. Piecework and fellowship piecework contributed substantially to increasing the amount and speed of production and therefore the risks in volume use of a highly toxic substance.[27]

The initial protection of women workers was to prevent explosion, and involved clean and dirty sides in workrooms, the donning of overalls, overshoes and caps, the body-searches to prevent forbidden matches or cigarettes entering the factory, the removal of all personal jewellery.[28] All these had an effect on the worker's consciousness of self and notion of her role as industrial worker, as a unit of production. By 1916, when TNT had become established as the major explosive in British armaments the welfare system had been set up by the Ministry of Munitions which ran the government factories and assisted the management in controlled factories. Welfare was originally aimed at women and the young both for external reasons, concerned with their social role and internal, as part of the management of production. The dominant motive was to keep the level of output rising. As the Gretna factory unit welfare supervisor summarised it,

> the welfare of the women operatives was considered
> by the Ministry and factory management as second
> only to the production of cordite.[29]

The worker on TNT found herself with two main agencies through which she could deal with TNT poisoning. One was the welfare system, the other workers' organisations — locally this meant the trade union. Despite the high degree of centralisation in the Ministry of Munitions and the persuasive argument of the official history of the

Ministry (that their welfare system was an all covering umbrella for war workers), recent evidence makes plain a wide variety of welfare procedures. In some places the welfare supervisor was well established, highly visible and supported by management — for example, the Royal Arsenal at Woolwich. In others, welfare supervisors were barely tolerated by management, ignored by most workers and unknown to many — for example, Armstrong Whitworth's in Newcastle. They were most effective in government factories where workers lived and worked in the same place — but this is not always true as a contrast between North Wales and Gretna shows.[30] In general welfare systems did not gain support from workers in all their activities — supervision outside the factory was condemned, body-screening resented and education classes unattended. Canteens and washrooms were popular as were football clubs, dances and choral societies.

The supervision of TNT workers to prevent poisoning seems to have been seen as an activity in which welfare supervisors were interested or responsible — not as an occasion for the self-activity of women workers. TNT poisoning met with very little organised or spontaneous resistance. Why was this?

One reason might have been the mutual suspicion of workers' organisations and welfare supervisors. Mary Macarthur, secretary of the National Federation of Women Workers (NFWW) and ex-secretary of the Women's Trade Union League (WTUL) was the leading speaker for women's trade unionism. Her union and other general unions enrolling women grew hugely 1915-1919. She said at a conference on welfare in 1918 that among women workers 'there is no word more hated than welfare'.[31] The Health of Munition Workers Committee issued emphatic instructions to welfare supervisors that they were part of management, not workers' representatives nor suppressors of trade unionism. This could, and did, happen. The Armstrong Whitworth's unit welfare supervisor reported that welfare supervisors:

> appeared to the workers in the light of spies who were going to watch and report to management . . . or as goody-goody people who were going to poke their noses into the workers' private affairs and interfere with their liberty and independence.[32]

However workers themselves often used the welfare system to express individual grievances. Several welfare supervisors reported excessive dependence by workers, one recorded frequent demands for wage-packet interpretation. Welfare workers at the lowest levels, the supervisors who gave medical checks, and the foremen who supervised the cloakrooms and did body-searches, would have been best placed to monitor severe deterioration in individual health. In practice it was left to illness to make the point — welfare workers concerned themselves most closely with good time-keeping. Unexplained absence from work or excess lateness could be caused by illness — and often it was not until the poisoning had made the worker so ill that she felt unable to work that it was detected.

Yet other factors in wartime affected women's time-keeping too — badly organised food supplies meant 2 hour queues in 1917; there were shortages of soap and coal so washing was hard; transport was overloaded and slow; there were few childcare facilities. Welfare did provide a rough and ready safety-net for the severely ill — but it did not do so in any consultation with the worker herself. This was a blunt instrument for dealing with TNT poisoning.

In its non-specific interest in the worker's health the welfare system reflected the views of women's organisations. The latter distrusted welfare's attempts to replace trade unionism but sought the *same* generalised protection for women's health as did the philanthropists who had developed welfare before the war. They exchanged personnel — both Isobel Sloan and Madeline Symons moved from the NFWW to goverment service. Trade unionists agitated for the protection of women's health in the interests of their national service — motherhood. A speaker at the 1919 TUC summed up the demand of mother's pensions

> If we have got to have an A1 nation we must protect
> the mothers. I honestly believe that the institution of
> pensions for mothers would go a long way towards
> checking the race suicide that is now going on.[33]

It is not simply that trade unions did not agitate over industrial disease, or poisoning. Lead poisoning had been the source of disquiet for many years, after a successful campaign it was made notifiable and a special sub-committee of the WTUL had been created to monitor its use and the incidence of poisoning. Lead poisoning was taken very seriously and attempts made in war-time to put women back into lead processes scrutinised.[34] The only explanation which seems plausible is that lead was known to be not only a killer but a major agent in gynaecological disorders and malfunction in childbearing. It was *also* not specific to war production.

A comparison between lead poisoning cases and TNT demonstrates the difference in severity in cases for women and men. Although of course the history of lead meant that 'cases' were more likely to be recognised and recorded while TNT poisoning did not necessarily lead to toxic jaundice.[35]

Toxic jaundice

Year	Men Cases	Men Deaths	Women Cases	Women Deaths
1916	48	23	122	34
1917	45	2	145	42
1918	7	2	27	8

Lead

Year	Men Cases	Men Deaths	Women Cases	Women Deaths
1916	318	20	30	1
1917	272	19	45	2
1918	124	19	20	0

Why did lead, less likely to cause death, lead to intense public agitation while TNT did not? The publications of the women's trade unions were silent on TNT except in 2 individual cases. They did protest about another toxic substance, aeroplane dope, which could cause death by accident through drowsiness as much as through jaundice. In 1917 dope was made safer by removing tetrachlorathane — its most toxic ingredient — from it.[36] TNT could not thus be rendered harmless. Their representatives did use TNT poisoning in argument in discussions at the Ministry of Munitions — but as an example of the increased exploitation of women which meant that they deserved war bonuses or rises in wages. Nationally they did not speak out on TNT poisoning at all — as far as can be seen in trade union press.

Locally of course it was different. Again though TNT was used as an example not as a cause — in one incident Addison wrote in his diary:

> I had to admonish a deputation from the Federation
> of Women Workers this afternoon. They have been
> holding meetings in Coventry. One of their speakers,

of an inventive turn of mind, has been giving a lurid
account of the dangers of TNT, saying, *inter alia*, that
all their inside organs would turn yellow and that
they would not be able to have children, etc.[37]

When seven girls refused to return to work on amatol (a compound
containing TNT) in 1917 they were presented at an industrial tribunal
and fined 15 shillings each. 'We are not labour conscripts', they said.
The NFWW commented cautiously in the *Woman Worker*, 'We do not
recommend our members to refuse work on explosives'.[38]

This fine, and many others for infringements of safety rules, was paid
by communal whip-round. It demonstrates the new self-confidence
among women workers, a new articulacy about their rights that was
notable in wartime and represented a general refusal to allow
militarisation of industry. But women's trade unions owed much of
their purchase on government to their power to discipline their own
members. To agitate against TNT in any wholesale way would have
been perceived as agitating against women's war work, logically
plausible after the pacifist statements of 1914 but technically
impossible after 1915 and their acceptance of dilution on terms
dictated by the trade unions of skilled men.

Oral evidence displays an ambivalent view among workers. Lack of
written record should not be taken as lack of concern. The women who
were yellow found the yellowing unpleasant. Some cafes would not
serve munition workers who added discoloration to the already low
status of 'factory girl' — they were supposed to be rough and ill-
mannered and were instantly detectable. One interview describes
officers in a first-class railway carriage looking at her 'as if we were
insects'. Another of her fellow workers said, 'They used just to frown on
the big factory because we were all yellow you see.'[39] One said very
poignantly,

> They called us canaries you know, but it wasn't nice
> like that, it was a horrible sickly green colour. My
> boys said, they wouldn't kiss me goodnight . . . oh
> mother, we don't like to see you green.[40]

Yet none of the interviewees recorded fright at the prospect of TNT
poisoning, only of accidents. The one worker who suffered any ill-
effects from TNT resented not being allowed back on the work.[41]
Another said she'd had TNT poisoning but described work on cordite
which she had chewed and had believed had had severe
gynaecological effects.[42]

This unconcern may reflect ignorance. The first deaths from TNT were publicised but were few; they could be seen as exceptional and blamed on the individuals concerned. In 1916, when most died, the information was suppressed. By 1917 though, in some factories regular checks were given and each girl had $\frac{1}{2}$ pint of milk a day which indicated to the workers that government was concerned about the problem, as did discipline over the use of gloves and respirators.

One of the women who had refused to work on amatol in 1917 said that she had a further grievance — that she should have been given a mask since 'doctors had argued against respirators, gloves and veils for CE work since they resulted in more dermatitis than without'.[43] These women were aware of the correct procedures and that management were not following them — but this perception did not result in any more generalised attempt to convey information to other workers; or an attempt to enlist others in the same cause. Evidence from recent interviews shows much greater fear of accident, particularly explosions, and greater experience of accident. The interview sample had some accidents — two broken legs and one broken arm because of bad lighting and poor workplace safety. One woman saw three severe accidents — one severed arm, one scalping, one crushed hand.[44] Accidents are quick, sudden and directly attributable to work, death from toxic jaundice was often slow and usually did not occur in public. But TNT was more complex an issue than even this contrast would imply.

The dominant notion in assessing perceptions of TNT work was the relation between war work and war service. Mrs Pankhurst changed the name of her march in 1915 which demanded a Women's War Register from the Right to Work march to the Right to Serve march — or so the discussion at the Ministry of Munitions would imply.[45] Women's work fell into the service category pre-war; women on explosives wore uniform; they enlisted for war service. The deaths from TNT were recorded on a 'Roll of Honour' so that death from an industrial disease was translated into a death on active service. [46] Several interviewees commented spontaneously, 'We didn't go through what the men did at the Front'. They certainly saw then and say now that the odds were very uneven — life expectancy at the Front was 6 weeks for much of 1915 and 1917. The ratio of dead to survivors of men on active military service was 1 to 7.5; while as far as can be calculated 0.1% of women TNT workers died.[47] The contrast was extreme and, although not quantified as such, was recognised by women workers and amid the rhetoric of sacrifice it is not surprising that women should have felt that their war service did not put them much at risk.

Secondly the war was a temporary phenomenon and so, therefore, seemed the production of TNT. It was a short period in a working life hence it could be controlled by short-term strategems rather than cured or rendered safe. Women did demand work after the war, but not the same work making instruments of death. For example, the prize-winning essay in a factory newsletter said of the writer's war-work:

> Only the fact that I am using my life's energy to destroy human souls gets on my nerves. Yet on the other hand, I am doing what I can to bring this horrible affair to an end. But once this war is over, never in creation will I do the same thing again.[48]

Although women's labour was in short supply 1916-1917, for the rest of the war there were enough workers to keep production rising and supply the Front — even if they had to be moved around from job to job.

The third reason for lack of any general systematic attempts to prevent TNT poisoning by its victims was that it could be prevented by individual systematic ways of dealing with war production — that is what management and government call labour turnover and absenteeism. The War Register of 1915 had turned all women into potential war-workers reducing a variety of labour histories into a pool of labour. The new factories run by the Ministry of Munitions were based on this source of labour and many women left areas of traditional female employment to work on munitions. Wages were much higher *relatively*. Women became more mobile and financially able to cope with short periods of unemployment. Witness after witness to the War Cabinet Committee on Women in Industry bemoaned women's bad time-keeping, lack of commitment in their work and lack of ambition as well as their high rate of turnover. They had an interest in so doing — showing that women were incapable of replacing men — but it is the case that this high mobility was seen as a problem of management.[49] It was not a problem for the workers themselves. From 1916-1918 women could move freely within factories and between factories — if free of domestic obligations. Since ultimately the only remedy for the build up of toxic material was to avoid taking in more, for the individual woman this was the most effective means of dealing with the problem. Change in the labour market and associated new independence among women were probably the major factors in reducing the incidence of TNT poisoning, as important as medical inspection and policing of the workforce in the interests of production.

A fourth factor in dealing with TNT poisoning was the other conditions affecting women's health. Some women's health improved — those on engineering processes or some skilled or clerical jobs in government factories benefited from higher wages than women earned before and some aspects of welfare. Munitions workers who ate in canteens got subsidised meat rations after 1917; TNT workers got half a pint of milk a day until 1917; simply eating away from home could mean a woman's first chance of eating a proper meal rather than what was left after others had eaten.[50] The Health of Munition Workers Committee's reports showed both improved stamina and output from attention to seating, light and lay-out as well as canteens, tea-breaks and medical attention.[51] The Ministry claimed that milk was ineffective as a specific against TNT poisoning. They dropped the daily half-pint in 1917 since milk was in short supply — but the general state of nutrition may have been improved by the protein intake.[52] Here the concern for motherhood allied to the need to keep production high and rising did not work unambiguously together. The Medical Research Council could find no *necessary* connection between nutritional state and resistance to TNT but felt the general improved level of nutrition among women could be justified as an indirect incentive to war production. It seems quite probable that more frequent investigation into TNT and its effects had the effect of revealing previously unrecorded conditions among young women in employment — anaemia for example could as equally have been attributed to war conditions as to TNT. The information we have is unreliable because of the short-term nature of the interest, the fears for morale and the mobility of the women themselves. The limitations of the state's knowledge of women's condition of health are shown by the nature of the discussion on TNT in Britain in 1914 to 1918. Oral evidence would also indicate a high level of deaths from workers who had been on explosives in the post-war influenza epidemic but the relationship needs closer examination to achieve any certainty.

The nature of the medical research and accounts of the effects of TNT were deeply affected by the job that the medical profession was being asked to carry out. There was no core of 'scientific medicine' which was to attain that 'perfect advancement and full knowledge which all desire'. Medical knowledge was treated as a distinctly different type of information — its power was used to disarm TNT workers who were faced with factory medical officers, medical inspectors and screening, milk and assurances that the problem was under control. To object to the effects of TNT on their health was made more difficult by censorship of medical and other information. Munitions workers did respond to TNT poisoning by refusal to do TNT work, by absenteeism and changing jobs — but they did not challenge the medical and scientific explanations. Their actions had to be based on individual experience combined with information and rumour passed by word of mouth. They did thereby succeed in forcing the Ministry of Munitions to protect their labour supply by taking action but beyond that they could act no farther. Their own organisations did not challenge the scientific explanations either despite their inadequacy and were more interested in the general conditions of women as mothers than these sufferers from a specific war-related disease. War provided a vast social laboratory for experiment on occupational disease and enabled doctors to claim TNT as a medical issue. This reduced the possible action of those most affected by the problem; emphasised the limited problem of death rather than the more extensive one of disease — and ensured that the lesson of industrial poisoning and its control should remain restricted to seeing it as a problem of production for doctors, management and the state, and not as a problem for the entire body politic.

Notes

1   W.H. Willcox, 'Lettsomian Lectures on Jaundice: with special reference to types occurring during the War', *British Medical Journal* (17 May 1919), p. 708, and Medical Research Council, *TNT Poisoning and the Fate of TNT in the Animal Body* , Special Report Series No.58 (HMSO, 1921), p. 5. (Henceforth *MRC*).

2   *History of the Ministry of Munitions,* unpublished, 1920-4 part III, 68.

3   Imperial War Museum, Women's Work Collection, Munitions 1^2 (IWM, Mun), Report of the Superintendent of HM Cordite Factory, Gretna.

4   Health of Munition Workers Committee Handbook, *Health of the Munition Worker* (HMSO, 1917), p. 97.

5   Sir George Newman, report in Addison papers, Box 2 at the Bodleian Library, Oxford.

6   Advisory committee minutes, Addison papers, Box 2 and PRO, MUN 4/1782.

7   *MRC* (1921), p. 25.

8   Notice to the Press, 1 Nov.1916, Addison papers, Box 2 TNT Advisory Committee instructions to the Press Bureau, 1916, PRO, MUN 4/1541

9   *MRC* (1921), p. 31.

10   A. Livingstone-Learmouth and B.M. Cunningham, 'Observations on the effects of tri-nitro-toluene on women workers', The *Lancet* (12 Aug. 1916), p. 261.

11   Ministry of Munitions, 'Trinitrotoluene Poisoning', The *Lancet,* (16 Dec. 1916), p. 1027.

12   *MRC* (1921).

13   Editorial, The *Lancet* (16 Dec. 1916), p. 1021, and Ministry of Munitions, 'Trinitrotoluene Poisoning', p. 1026.

14   Dr. W.J. O'Donovan, 'Circular to Medical Officer in filling Factories', 1916, Addison papers, Box 2, and R.H. Gummer, *Barnbow no.1 (Leeds) National Filling Factory,* unpublished (n.d.), p. 45.

15   *MRC* (1921), p. 16.

16   Health of Munition Workers Committee, *Final Report, Industrial Health and Efficiency* (1918), Cd 9065, p. 78.

17   B. Moore, *BMJ* (4 Aug. 1917), p. 164.

18   MRC (Committee) Special Report Series no.11, *The Causation and Prevention of Tri-Nitro-Toluene (TNT) Poisoning,* (HMSO 1917), p. 47.

19   Ministry of Munitions, 'Trinitrotoluene Poisoning', *BMJ* (Dec. 1916), p. 844.

20   The *Lancet* (16 Dec. 1917).

21   Minutes of the TNT Advisory Committee, PRO MUN 4/1782.

22   Ministry of Munitions, issues to medical officers in filling factories. 1916, Addison papers, Box 2.

23   PRO, Mun 4/4872.

24   *MRC* (1917), p. 59.

25   Recruitment posters, Imperial War Museum

26   *Labour Gazette* (Dec. 1917), p. 438.

27   G.D.H. Cole, *Trade Unionism and Munitions* (Oxford, 1923).

28   Health of Munition Workers Committee, Memo no.4, *The Employment of Women* (1916), p. xxiii, Cd 8185.

29   IWM. Mun 14, Report from National Shell Factory Gretna, p. 9.

30 D. Thom, unpub. Ph.D., The Ideology of Women's Work, 1914-1924; with special reference to the NFWW and other trade unions, 1982, CNAA, chap.6.

31 cit B. Webb in K. Dewer (ed.), *The Crisis* (1920).

32 IWM, Mun 19, Armstrong-Whitworth's.

33 The Congress Report, 1919 cit G. Braybon, *Women Workers in the First World War* (London, 1981), p. 199.

34 *Women's Trade Union Review.*

35 A. Anderson, *Women in the Factory* (London, 1922), p. 307. IWM, Mun 34^2, Roll of Honour (those killed on Munitions Work) This records 76 deaths from TNT poisoning.

36 PRO, MUN 2/27, 16 Sept 1916.

37 Addison diary. Addison papers.

38 *Woolwich Pioneer* (16 Feb. 1917); *The Woman Worker* (March, 1917);

39 IWM Recording. C. Renolds 000566/07, Reel 1. IWM Recording. E. McIntyre 000673/09, Reel 1.

40 Interview Mrs. L. Robinson, July 1977, D. Thom.

41 Interview Mrs. MacKenzie, June 1977, D. Thom.

42 Interview Mrs. Cushin, June 1977, D. Thom.

43 cf 38. Interview Mrs. Bennett, July 1977, D. Thom.

44 Interview Mrs. L. Robinson, July 1977. D. Thom.

45 PRO, Mun 5.70 11 Aug. 1915, 28 Aug. 1915.

46 IWM, Mun 34^2.

47 Calculation based on figures from N.B. Dearle, *The Cost of the War* (Newhaven, 1924), My thanks to Dr. Jay Winter for this reference.

48 IWM, Mun 28, Alexandria Filling Factory.

49 *Report* of the War Cabinet Committee on Women in Industry, pp. 1919, Cmd 135, and *Evidence* to the War Cabinet Committee on Women in Industry, Cmd 167.

50 Mrs Pember Reeves, *Round About a Pound a Week* (London, 1911), demonstrates the prewar eating habits of mothers in low-income households.

51 Health of Munition Workers Committee Handbook, *The Employment of Women* (1917).

52 PRO, MUN 2/28, 27 April 1 p. 918,9.

6 TUBERCULOSIS, SILICOSIS, AND THE SLATE INDUSTRY IN NORTH WALES 1927–1939

Linda Bryder

The slate industry had become a commercial enterprise in the counties of Caernarvonshire and Merionethshire in North Wales by the mid-eighteenth century.[1] At the end of the nineteenth century, this area included the two largest slate quarries in the world, Penrhyn Quarry near Bethesda and the Dinorwic Quarry at Llanberis, and the largest slate mine, Oakeley at Blaenau Ffestiniog, as well as fifty smaller mines and quarries scattered on the hillsides.[2] The industry was at its height at this time, employing 13-15,000 workers. Overseas competition and changing fashions in roofing adversely affected the industry in the early twentieth century and it was never to recover its former prosperity. In 1910 there were still 13,000 workers in the industry, but by 1945 the number had dropped to 3,520.[3] A recent study of the health of the workers in the industry by J. R. Glover *et al.*, published in 1980, showed pneumoconiosis to be very prevalent.[4] Tuberculosis has not been a major problem in that industry or elsewhere in Britain since the early 1950s when streptomycin and related drugs were introduced. However, Glover's study showed many of the lungs of the older miners to have healed tuberculous lesions, indicating a very high prevalence of tuberculosis among workers in the industry some thirty years previously. Modern epidemiological studies suggest that silicosis and pneumoconiosis predispose to tuberculosis.[5]

The first detailed study of the health of the North Wales slate districts was published by Dr T. W. Wade of the Welsh Board of Health in 1927.[6] This was followed by a survey by Drs C. H. Sutherland and S. Bryson of the Mines Department in 1930,[7] and by Dr H.D. Chalke of the King Edward VII Welsh National Memorial Association in 1933.[8] A Ministry of Health Committee of Inquiry into the Anti-Tuberculosis Service in Wales and Monmouthshire, set up in 1937 under the chairmanship of Clement Davies, Liberal MP for Montgomeryshire and future leader of the Liberal Party, which reported in 1939, also received evidence from the area in their investigations.[9] The causes of tuberculosis in North Wales were being discussed constantly throughout this period in the press and at public meetings. This chapter considers why attention was directed to the health of this area, and focuses on the discussions of the local medical profession, their recommendations, and the underlying assumptions concerning health and disease which the discussions revealed. The attitudes of the workers themselves are also considered as far as they can be ascertained.

It was not interest in silicosis, or the hazards of the slate industry, which drew attention to the health of the North Wales quarrying districts, but rather the high tuberculosis death rates which prevailed in the area. This concern for the high tuberculosis rates was partly related to the general interest evidenced in the state of the nation's health during the inter-war period, also reflected in studies of malnutrition and poverty. Tuberculosis was often said to be a reliable index of the health of the people, and its causes to lie in social conditions. Richard Titmuss, for example, wrote in 1939,

> This disease, of all those studied, appears to be the
> most sensitive to variations in such indices of poverty
> as unemployment allowances, poor relief and a
> sustained experience of depression in a given area. . .
> Almost always such indices are faithfully reflected in
> a rise of tuberculosis mortality and morbidity,
> especially in the responsive age groups (15-35) for
> both men and women.[10]

The Chief Medical Officer, Sir George Newman, also wrote in 1921, 'The close association of poverty or lack of adequate nutrition with a tendency to higher tuberculosis rates is increasingly evident.'[11] However, such statements, increasingly charged with political implications,[12] were to become less frequent in the medical officer's reports.

An investigation into one of the 'black spots' of tuberculosis, Tyneside was carried out in 1933.[13] Attention was also drawn to Wales as possessing some of the worst 'black spots' of tuberculosis, particularly North Wales. While the tuberculosis death rate for England and Wales from 1930-6 was 0.724 per thousand population, those for the slate quarrying districts of Caernarvonshire, Gwyrfai Rural District and Pwllheli Borough, were 2.052 and 1.718 respectively.[14] Moreover, while the national death rates had been declining, the rates for males in Gwyrfai had risen from 1.88 per thousand in 1909-13 to 2.16 in 1921-5, and to 2.43 in the period 1922-31.[15] In Merionethshire, the slate district of Ffestiniog had a tuberculosis death rate of 1.48 in 1922-31 and 1.938 in 1930-36.[16] The

two counties of Caernarvonshire and Merionethshire had the highest
death rates from tuberculosis of all counties in England and Wales in
1932.[17]

Another important factor in fostering the interest in studies of
tuberculosis was the very existence of organisations and offices for
dealing specifically with tuberculosis, which had been growing up
since the early twentieth century. In the nineteenth century those who
contracted tuberculosis had been largely ignored or relegated to Poor
Law institutions; tuberculosis had been regarded as an inherited,
constitutional disease over which medicine had little control. A change
of outlook followed the establishment of a special type of tuberculosis
institution in Germany by Hermann Brehmer in 1859. Brehmer had
argued that tuberculosis could be cured and claimed successful results
from the 'open-air' treatment practised in his institution. It was not
until the 1890s however that the movement spread to Britain. An
important factor was the discovery of the tubercle bacillus (or
Micobacterium tuberculosis) the causal agent of tuberculosis, by Robert
Koch in 1882. This discovery brought tuberculosis into line with other
infectious diseases and led to a more positive approach to prevention
and cure. Despite Koch's claim to have discovered a cure in tuberculin
(a culture of the tubercle bacillus), no actual cure had yet been found.
Nevertheless, there was great enthusiasm for 'open-air' or 'sanatorium'
treatment. Moreover, there was a belief among the medical profession
that, now that the cause was known, the discovery of a cure was
imminent, and hence an enthusiasm for research which could best be
undertaken in institutions. Others saw the value of institutions in that
they isolated the source of infection, and yet others that they taught
patients how to control their infectiousness. Thus, in 1886 there had
been nineteen hospitals specialising in tuberculosis in England and
Wales;[18] by 1920 there were 388 institutions providing for tuberculosis
patients as well as 398 dispensaries.[19] These in turn generated a body
of specialists whose particular concern was the treatment of and
research into tuberculosis. The National Association for the
Prevention of Tuberculosis (N.A.P.T.), founded in 1898 (which
financed the 1933 study of Tyneside) was one manifestation of the new
interest in the disease, and the King Edward VII Welsh National
Memorial Association, founded in Wales in 1910, was another.[20] The
latter was specifically founded to provide institutional treatment of
tuberculosis in Wales, but became increasingly involved in other
aspects of the tuberculosis problem, that is in prevention and research.
It was responsible for the 1933 study of Gwyrfai. It was also reponsible

for initiating the 1937-39 Inquiry into the Anti-Tuberculosis Service in Wales and Monmouthshire, as the result of a dispute with the Welsh Local Authorities over, the latter's financial contributions to the services. By the 1921 Public Health (Tuberculosis) Act, institutional treatment of tuberculosis had become a statutory obligation of local authorities. A special clause had stipulated that Welsh authorities would be fulfilling their obligation by coming to a financial arrangement with the Memorial Association. The amount demanded by the Association for tuberculosis services had steadily increased until 1937 when their estimated expenditure for 1941 amounted to £380,000.00 (annual costs for 1930-33 had been £206,361.00).[21] The Welsh Local Authorities objected to the increase claiming they were spending a disproportionate amount of their public health finances on tuberculosis. A representative of the Flintshire County Council pointed out that the Association swallowed up fifty per cent of the County's gross expenditure on health services: 'Having regard to the fact that tuberculosis is responsible for only 7% of the total deaths in Wales, it seems to us that this expenditure is out of proportion.'[22] The Association argued that the increased expenditure was necessary and justified, and they demanded an inquiry into the tuberculosis services in Wales.[23]

Thus it was tuberculosis in the general population and not specifically diseases related to the slate industry which attracted attention to the area, although some link between the industrial process and tuberculosis had been suspected as early as the late nineteenth century.[24]

Dr Wade was sent to investigate the area in 1926 by Neville Chamberlain, then Minister of Health, in response to a question in the House of Commons by Major Lloyd Owen, Member of Parliament and Medical Officer for Caernarvonshire (Southern Division), concerning the excessive tuberculosis rates in the area.[25] Wade's analysis of slate dust showed up to fifty per cent quartz content.[26] The quartz particles in the slate dust produced silicon dioxide which, when inhaled, irritated the lung and caused excessive build-up of tissue. Wade concluded that inhalation of slate dust was causing silicosis among the workers which in turn predisposing them to tuberculosis. He concluded furthermore that, contrary to popular belief at the time, the tuberculosis of quarrymen was not less infectious than among the general population, and that therefore the quarrymen were an important source of infection for the other residents in the districts. The industrial process was, in his opinion, an important factor in the high tuberculosis death rates in the area.[27]

Tuberculosis was not a scheduled disease under the Workmen's Compensation Acts, but silicosis had been included under the Acts since 1918. In 1918 only ganister miners and makers of silica brick who suffered death or total disablement were included.[28] This was broadened in the 1920s to include other industries, but not the slate industry. Nor did Wade's report result in the inclusion of the industry in the Silicosis Scheme under the Acts. The *Western Mail* reported that 'the investigation carried on in an approved scientific manner appears to show that the real cause [of the high tuberculosis rates in the area] is the inhalation and swallowing of fine slate dust...'[29] However, the majority of the medical profession practising in the area did not accept Wade's results.[30] One doctor he interviewed was definitely of the opinion that slate dust was not harmful but beneficial.[31] Nor did the quarry managers whom Wade interviewed agree with his results, but were emphatic that slate dust was not injurious to the workers; one cited the large number of old workmen who still carried on their work in the industry as evidence.[32] It was reported in the *Slate Trade Gazette* in 1927 that,

> There is hardly any escape from the inhalation of dust by those who rely on industry for a livelihood...
> The probability is that road dust is just as injurious to the human system as slate dust for it contains minute particles and it is almost certain that slate dust is not so harmful as that teaming with malignant bacteria. It is all a question of relativity and opinions differ widely on the subject. That tuberculosis is very prevalent in North Wales may be a coincidence... Everyone wishes to give the workers the best possible conditions but if all sorts of fantastic rules and regulations, incurring considerable expense, are to be foisted upon employers, the time will come when the businesses of the latter will not be worth continuing.[33]

They concluded with the warning that continued impositions and interferences by the state meant the diminution of capital and more unemployment.[34]

Nor did a subsequent investigation instigated by the Mines Department support Wade's findings. In their investigation, Drs Sutherland and Bryson examined 120 men: 56 of whom showed evidence of fibrosis, 14 of silicosis, and only 3 of simple tuberculosis.

They concluded that, 'From this it appears that the industry is not one that renders the workmen peculiarly liable to contract pulmonary tuberculosis,' [35] a conclusion which, according to Wade and Dr C. Dairel, a tuberculosis physician from Cefn Mably Tuberculosis Hospital near Cardiff, went further than the evidence warranted.[36] The latter pointed out that of the 120 cases examined, only 61 were examined by the use of an x-ray which he considered essential.[37]

The King Edward VII Welsh National Memorial Association also became involved in the debate on the causes of the excessive tuberculosis death rates in this area. Research was an increasingly important part of the work of this Association. Attention was drawn at a Council meeting in 1931 to the high tuberculosis rates in certain areas, which persisted despite the work of the Association.[38] S. Lyle Cummins, Professor of Tuberculosis and Consultant to the Association, prepared a memorandum on the subject.[39] He referred to Wade's report but did not discuss slate dust as an important predisposing cause. He nevertheless pointed to the necessity of further research in the area. Dr H.D. Chalke, Assistant Tuberculosis Officer for West Monmouthshire, was appointed to carry out this research into Gwyrfai, a 'black spot' of tuberculosis. The Ffestiniog Council also invited the Association to include their district in the survey following pressure from the Medical Officer for Ffestiniog Urban District, Dr J.W. Morris, who believed that slate dust predisposed quarrymen to tuberculosis.[40]

However, for those hoping for a definitive statement on the influence of dust, Chalke's report, published in 1933, must have been a disappointment. He referred to Wade's report, but remained uncommitted on the dust issue. He wrote, 'Medical opinion in Gwyrfai and other parts of North Wales seems to differ considerably as to the association of dust-inhalation and phthisis [pulmonary tuberculosis], the weight of opinion tending to discredit such an association.' [41]

In the discussions on the causes of tuberculosis in the area which followed the publication of Chalke's report, dust inhalation in the slate industry was rarely mentioned. When it was remarked upon, it was to the effect that it had not been proved that slate dust was conducive to a high death rate among quarrymen.[42] One suggestion was that the money spent on dust allaying experiments, limited though these were,[43] might be better employed in providing facilities to ease the strain placed upon the workmen in carrying out the heavy part of their work,[44] in accordance with the theory that strain was an important causal factor in tuberculosis.

The 1937-9 Committee of Inquiry into the Anti-Tuberculosis Services in Wales and Monmouthshire also investigated the area and came to the conclusion, on the evidence of the Tuberculosis Officer for Ffestiniog, Dr T. Watkin Davies, and the Medical Officer, Dr J. W. Morris,[45] but contrary to the belief of the majority of their medical witnesses, that inhalation of slate dust caused silicosis and was an important factor in the excessive tuberculosis rates of the area. The Chairman of the Committee, Clement Davies, expressed surprise that the workers were not eligible for compensation and wrote to the Home Office to urge their inclusion under the Silicosis Scheme.[46] Watkin Davies also presented the results of his work in the area (which included an x-ray examination of 117 cases) to a meeting of the Tuberculosis Association in 1939 at which Dr E. L. Middleton of the Home Office was present.[47] Following the publication of Davies' evidence in 1939,[48] slate miners were included in the Silicosis Scheme of the Workmen's Compensation Act.[49]

If Wade and Davies were not representative of the current medical views of the 1930s, then to what did the majority attribute the high tuberculosis death rates? Wade had also isolated economic conditions as an important factor and believed that an improvement would result from a higher living standard.[50] Moreover, as already pointed out, it was interest in tuberculosis as a 'social' disease which was attracting much public attention to it at this time. Dr Chalke stressed sanitary and hygienic defects and was highly critical of local government. According to *The Times*, 19 June 1933, the conclusions of Chalke's report

> cannot fail to be highly disturbing to those concerned
> in the administration of public health in Wales. The
> evidence respecting sanitary conditions in some of
> the villages in the slate-quarrying districts of
> Caernarvonshire is startling in its revelation of a state
> of affairs which would seem to be impossible in any
> part of the country.[51]

However, the social and economic origins of the disease were not generally among the factors dwelt upon in the discussions on North Wales in the 1930s. Attention was focused on the following factors: race, family, heredity, the fatalistic attitude of the people, improvidence and inefficiency of housewives, particularly as regard to diet, and the general social habits of the workers and their families.

In his 1930 memorandum on the high tuberculosis death rates in certain parts of North Wales, Professor Lyle Cummins cited the work of Dr Emrys Bowen, Cecil Prosser Research Scholar in tuberculosis at the University of Wales, who explained the differences between various districts in Wales in terms of race; the dark long-headed type was apparently more easily adaptable to industrial environment than the fair-haired Anglo-Saxon.[52] Dr W. H. Lewis of the Montgomeryshire Insurance Committee believed that tuberculosis was more prevalent in Western Wales because the Iberian type of Welshman was to be found there.[53] J. E. Tomley, clerk to the Montgomeryshire Insurance Committee and member of the Council of the Association, who was largely instrumental in drawing the attention of the latter to the need for an investigation in the area, regarded the coincidence of tuberculosis and Welsh speakers as important.[54] Dr A. C. Watkin, Tuberculosis Officer for Salop County Council, also claimed at a meeting of the N.A.P.T. in 1933 that 'in the same counties where you find tuberculosis lingering in this acute form you also find the Welsh language surviving most strongly.'[55]

Discussing Chalke's 1933 report, an article in the *Liverpool Post and Mercury* pointed out that it had already been agreed as a matter of fact that the absence of a definite and satisfactory explanation of the extreme prevalence of the disease in Gwyrfai more than in any other area was a disappointing feature in an otherwise admirable report. The article quoted 'one professionally interested student of the subject, a man who had lived and worked with the people all his life, and whose opinion was sought for the purpose of the report'. This student apparently stated emphatically that no amount of new housing, sanitation, or education would remove the scourge. The root cause, he believed, was sociological, and it was not so much new houses, new food and new ideas that the people required as new blood. Inter-marriage and inbreeding had been practised down the generations to such a degree of complexity that, in his opinion, 'the family trees in the affected villages have not only run riot but have run to seed'.[56] Chalke had started to plot family infection, as Dr R. C. Hutchinson had done in a survey in Carmarthenshire,[57] but had given up the attempt when it seemed that the six or seven hundred people in one village all belonged to one of three families.[58]

Giving evidence to the 1937-39 Committee of Inquiry, one doctor from Gwyrfai Rural District said, 'Contrary to all that I was taught I have never seen a case of tuberculosis without some family history... To me it seems that these people are born without the vital resistance to combat tuberculosis', and he advocated the notification of the family as a whole and that their children should receive special treatment from birth, (although the nature of this special treatment remained unspecified).[59] Another doctor from the area also referred to inter-marriage among members of tuberculous families as leading to increased tuberculosis in this area.[60] In 1936 the Caernavonshire Joint Sanitary Authority recommended a voluntary examination of each prospective partner before entering matrimony.[61] Marriages of near relations, they said, were to be deprecated, especially when the same weakness was present on both sides. Legislation prohibiting the marriage of tuberculous persons was not introduced into this country as it was in Germany in 1936, although it seems that at least some would have been in favour of it. Despite the discovery of the tubercle bacillus, and the infectious nature of the disease, by Robert Koch in 1882, belief in the heredity of tuberculosis persisted in the form of tuberculus diathesis.

Another factor under discussion was the supposed natural fatalism of the people, and their refusal to undergo institutional treatment. Chalke maintained 'This failure to face facts, combined with a rather fatalistic outlook, makes the control of tuberculosis very difficult'.[62] Dr V. Emrys Jones, Tuberculosis Officer for Anglesey and Caernarvonshire, also said of the inhabitants of North West Wales in 1938 that they were mainly a cultivated, well-read and intelligent people, even in many instances in poorer classes but he said they were imaginative, sentimental and highly fatalistic, and regarded tuberculosis more as a disgrace than a disease, which made treatment difficult.[63] Tomley mentioned the refusal of the people to open their doors to the tuberculosis officer because of the stigma attached to the disease.[64] According to the *Cambrian News and Welsh Farmers Gazette* , 'Even mentioning the name of the tuberculosis physician was sufficient to frighten some persons into saying 'no' [to examination]'.[65] Thus it was said they did not come under institutional treatment until it was too late. There was, however, no mention of the long waiting lists for institutional treatment which existed at this time,[66] nor of the financial and social consequences of the discovery of tuberculosis for the person concerned and his or her family, a very practical reason for not seeking advice and treatment at a time when the only financial assistance in such cases was charity or the Poor Law. Nor was there any questioning of the efficacy of institutional treatment in curing those who did undertake it.[67] Moreover, this fatalistic response to tuberculosis was not peculiar to the inhabitants of Wales; tuberculosis officers in England were reporting the same difficulties.[68]

An association between tuberculosis and nutrition had frequently been noted, especially following World War I when tuberculosis rates were shown to have increased dramatically in areas with restricted food consumption.[69] Medical opinion in North Wales also isolated diet as an important factor in tuberculosis causation. The doctors whom Wade interviewed in 1927 considered this to be the most important factor, but they believed the inefficiency and thriftlessness of the quarrymen's wives were largely responsible. Dr Griffiths of Bethesda maintained that the quarrymen's wives were thriftless and did not prepare meals properly for their menfolk who had to subsist too much on tea, bread, butter and tinned foods.[70] Dr John Roberts of Llanberis believed that the quarryman's mode of living and poor feeding on tea, bread, butter and over-eating on Sundays reduced his powers of resistance.[71] Chalke also studied the diet of the people and came to the conclusion that 'The diet of the quarryman is unjudicious rather than insufficient'.[72]

The educational system was generally blamed for the deficiencies in the quarrymen's domestic economy. At an annual meeting of the Welsh National Memorial Association, Reverend H. R. Protheroe of Bridgend stated that the educational authorities were packing the minds of their young women at the expense of their bodies and impairing their constitutions for the rest of their lives.[73] The Principal Medical Officer of the Association, Dr D.A. Powell, agreed with this to an extent. He maintained that eighty per cent of those young women would end up as wives and that the next generation would be a sounder one if they were taken away from the academic side and turned to domestic work. It was pointed out at this meeting that women married young with little idea of how to care for their families. All they apparently knew was the use of a tin-opener and a cork screw.[74]

Dr Llewelyn Williams, Senior Medical Officer of the Welsh Board of Health, believed that it was about time they turned back to the diet of their forefathers instead of the present tinned food.[75] Dr John Jones of Dolgellau agreed; the people in these days did not eat proper food.[76] Thus there was a nostalgia for the mythical 'good old days', when the yeoman farmer tilled his own land and produced his own food.

The 1937-39 Committee of Inquiry received evidence along the same lines. Dr Norris of Neath believed a certain amount of undernourishment existed in Wales, mainly due to ignorance and bad housekeeping.[77] Dr Hawkins and Dr Rowland Williams of Pembroke believed malnutrition in the majority of cases was due to ignorance.[78] Dr Roberts of Flint maintained that incorrect feeding rather than poverty was often responsible for malnutrition and that too much use was made of the tin-opener and synthetic foods.[79] The Committee concluded, 'The tendency of housewives to neglect, through ignorance of dietetic values or in order to save trouble, the old traditional forms of feeding and to rely largely on prepared foods is considered to have some bearing [on the high tuberculosis rates].'[80] Chalke, however, believed that 'bad tradition' had existed for centuries.[81] A 1893 committee of inquiry had shown that the workers lived at that time mainly on bread, butter and tea.[82]

Other social habits of the workers and their families were also commented on. Chalke did not discover alcoholism to be prevalent in the area and therefore a contributory factor in the high tuberculosis rates. He considered the religious principles of the people to be an important factor, for it resulted in their congregation in ill-ventilated, overcrowded churches. Another factor, he believed, was the lack of opportunity for suitable recreation in the open air. Garden shelters, introduced by tuberculosis dispensaries for domiciliary treatment, were said to be conspicuously absent in Gwyrfai, and the value of open windows was apparently not fully appreciated. The danger of kissing, he said, was too often disregarded, and the young quarrymen took a great pride in their personal appearance, particularly on Sundays, leading to a tendency to stint themselves to spend more on clothing.[83]

A medical witness before the 1937-9 Committee also stressed the great importance attached to 'turning out smartly'; instead of dressing serviceably to suit weather conditions it was too often done with the object of 'equalling or bettering the dress of their friends'.[84] This witness deprecated what he said was common knowledge among those who were acquainted with the slate quarry districts, that is the tendency for the family to congregate in the back kitchen while a more commodious room was only used as a 'showroom' on auspicious occasions. He also referred to the 'too common practice' of excluding sunshine with blinds and heavy curtains, and the reducing of airspace by overcrowding rooms with furniture. The heavy local rainfall was considered a contributory factor, for many of the quarrymen often worked in wet clothes with resultant high incidence of colds.[85]

Thus the medical commentators placed the emphasis neither on the slate dust nor economic and social conditions, but on the personal habits of the workers and their families. The people themselves were seen to be responsible for the state of their health. Some considered tuberculosis to be hereditary and regarded inter-marriage as important, others blamed the social customs of the people, their fatalistic attitude towards the disease, an improper rather than inadequate diet, as well as insufficient sleep, fresh air, and outdoor recreation. This perception of the problem was not, however, unique to the Welsh. Sir John Robertson, Medical Officer of Health for Birmingham and Professor of Public Health at the University of Birmingham, for example, explained the differences in tuberculosis death rates between poorer and better-class districts of Birmingham in terms of the 'ignorance and carelessness of the inhabitants', and believed it to be wonderful that more young people did not contract tuberculosis, 'for so large a number are living unhealthy lives in one direction or another.'[86] Self-responsibility was also the dominant theme of the health propaganda of the N.A.P.T. The conclusion of the N.A.P.T. film, ' The Invisible Enemy' : 'Go and teach the truth. The fate of each man is in his own hands,' sums up the ideological position of the N.A.P.T.[87]

How did their perception of the problem affect the responses of the medical profession? Not seeing the problem in political terms, they did not seek political remedies; on the contrary, their medical views served to support the establishment. Among their recommendations, measures such as improved conditions in the mines and quarries, higher wages, and compensation did not figure prominently.

Little was done to allay dust hazards in the 1930s. E. Andrewes, the Managing Director of Maen Offeren Slate Quarry Company, told the Committee of Inquiry in 1937,

> Respirators are now, I believe, provided in all the Ffestiniog Slate Quarries, but in any case there is nothing to prevent the men providing themselves with such appliances. A quite efficient respirator can be purchased for the sum of 3/6d. I am told however that many of the experienced men prefer tying a handkerchief over the mouth and nose.[88]

The local Medical Officer, J.W. Morris, explained however that, in the disease of silicosis, one of the persistent features was the difficulty of breathing; when the quarrymen put on the mask the difficulty was increased, so they would not wear them.[89] Dr Lloyd Owen, MP for Caernarvonshire, expressed regret in 1937 that the quarries in the Gwyrfai district had not made much progress in the matter of allaying dust.[90]

The solution most commonly recommended by the medical commentators was education, of housewives in particular. Workers and their families were to be taught to make the best of existing circumstances. Dr D.A. Powell, Medical Superintendent of the North Wales Sanatorium had already introduced a widely commended scheme for training female patients in housework while resident at the sanatorium. They were placed in cottages 'such as would make a sanitary inspector weep', 'equipped with every drawback, crammed with furniture and replete with all the gadgets beloved of the sanitary defective'. The patients were to be taught how to make such cottages habitable.[91]

Following Chalke's report the Welsh National Memorial Association drew up an exemplary budget for housewives, and appointed two health visitors in the area to visit the homes and advise on domestic economy.[92]

Workers themselves were conspicuous throughout the 1930s for their silence on the question of the causes of tuberculosis and the influence of slate dust. Samuel Hoare, the Home Secretary, wrote in reply to a letter from Clement Davies in 1939 that he believed that the slate workers could possibly apply for compensation under special clauses of the 1931 Silicosis Scheme.[93] However, not a single claim had been made. The procedure for claiming compensation was expensive and conditions of eligibility far from clear. The general understanding of the situation revealed in the 1937-9 Inquiry was that the workers were not eligible for compensation because they did not come under the 1930 Silicosis (Workmen's Compensation) Act.[94] Nor was there any attempt by the workers or their union in the 1930s to alter the situation. Their silence may have been related to the fatalism noted by investigators. Possibly a more important reason for their silence was the fear of the economic and social consequences of contracting tuberculosis or being discovered to be tuberculous. A full-scale investigation would lead to instant dismissal of the tuberculosis cases discovered to prevent further spread of the disease among employees. Fear of loss of employment should not be underestimated given the economic climate of the 1930s and the welfare provision available, or lack thereof. Moreover, the employers did not hesitate to point out that large compensation bills might result in the closure of quarries through bankruptcy. The slate industry was not a thriving enterprise at any time in the twentieth century, and was in a state of constant decline. As already pointed out, 13,000 workers were employed in 1910; by 1945, 3,520.[95] Even the Secretary of the North Wales Quarrymen's Union, R. T. Jones, told Wade in 1927 that he was not anxious that any additional burden should be placed upon the industry in the way of compensation to affected workmen.[96] Action on compensation was to come initially from outside rather than from the workers themselves, specifically from the 1937-9 Committee of Inquiry and an officer of the Welsh National Memorial Association.

Thus attention was drawn to the North Wales slate quarrying districts as a 'black spot' of tuberculosis. The interest was partly related to a broader concern for tuberculosis as a 'social' disease, which must be seen in the context of the social surveys of the inter-war period, and partly to the existence of a body of specialists in tuberculosis which had been growing up since the early twentieth century. The local medical profession did not subscribe to a doctrine of the social and economic origins of the disease, nor did they consider industrial processes to be important. Their views were clearly located in the dominant ideology of self-help; their aim was reformation of the personal habits of the people. While their opinions reflected dominant ideology, they had greater weight than those of politicians and journalists, for they were presented as part of a scientifically-based medical discourse. As the sociologists Wright and Treacher point out, medicine had a privileged epistemological status that was usually accorded to science: if science was the accurate reading of Nature, undistorted by social interest or cultural bias, then medicine could claim to be the benevolent application of what was found in the natural world.[97] Whether they succeeded in changing the habits of the workers in the industry and their families is unknown; what is clear however is that the latter did not dispute the views promulgated by the medical profession. The response of the workers to the debate on the high tuberculosis rates in the industry and the area seems to have been dominated above all by fear — that fear of the economic and social consequences of being found to be tuberculous, described by Dr Chalke as a 'failure to face facts'. [98]

Notes

1   For the most recent histories of the industry and the area see Jean Lindsay, *A History of the North Wales Slate Industry* (Newton Abbot, 1974); R. Merfyn Jones, 'Y Chwarelwyr: the Slate Quarrymen of North Wales', in Raphael Samuel (ed.), *Miners, Quarrymen and Saltworkers* (London, 1977); and R. Merfyn Jones, *Studies in Welsh History 4. The North Wales Quarrymen 1874-1922* (Cardiff, 1982).

2   R. Merfyn Jones, *The North Wales Quarrymen*, p. 72; 'Y Chwarelwyr', p. 101.

3   Jean Lindsay, *Slate Industry*, p. 298.

4   J. R. Glover et al ., 'Effects of exposure to slate dust in North Wales', *British Journal of Industrial Medicine*, vol. 37 (1980), pp. 152-160. 'Pneumoconiosis' is used in his study as a more accurate description than 'silicosis', although the distinction had not yet been made in the period this chapter is concerned with.

5   Ibid. Also see Alice Stewart, 'Tuberculosis in Industry', F.R.G. Heaf (ed.), *Symposium of Tuberculosis* (London, 1957), pp. 645-84. Walter Pagel et al., *Pulmonary Tuberculosis: Bacteriology, Pathology, Diagnosis, Management, Epidemiology, and Prevention*, 4th edn (Oxford, 1964), p. 66.

6   *Reports on Public Health and Medical Subjects No. 38A, Report of an Investigation into the Alleged High Mortality Rate from Tuberculosis of the Respiratory System among Slate Quarrymen and Slate Workers in the Gwyrfai District*, T. W. Wade (Welsh Board of Health, London, HMSO, 1927).

7   *Report of an Inquiry into the Occurrence of Disease of the Lungs from Dust Inhalation in the Slate Industry in the Gwyrfai District*, Drs C. H. Sutherland and S. Bryson (Mines Department, London, HMSO, 1930).

8   Dr. H. D. Chalke, *Report of an Investigation into the Causes of the Continued High Death Rate from Tuberculosis in Certain Parts of North Wales* (King Edward VII Welsh National Memorial Association, Cardiff, 1933).

9   *Ministry of Health, Report of the Committee of Inquiry into the Anti-Tuberculosis Service in Wales and Monmouthshire (Clement Davies Report)* (London, HMSO, 1939).

10   Richard Titmuss, *Poverty and Population, a Factual Study of Contemporary Social Waste* (London, 1938), pp. 169, 170.

11   Sir George Newman, *On the State of Public Health, Annual Report of Chief Medical Officer to the Ministry of Health for 1921* (London, HMSO, 1922), p. 62.

12   C. Webster, 'Healthy or Hungry 30s?', *History Workshop Journal*, no. 13 (1982), pp. 110-129.

13   F.C.S. Bradbury, *Causal Factors in Tuberculosis* (National Association for the Prevention of Tuberculosis (N.A.P.T), 1933).

14   *Clement Davies Report*, p. 258.

15   *Wade Report*, p. 6; *Chalke Report*, p.20.

16   *Clement Davies Report*, p.259; *Chalke Report*, p.68.

17   *Clement Davies Report*, pp. 258-9; Public Records Office, Ministry of Health files, MH96/1111, 'Mortality from Tuberculosis 1932 and 1933 in Administrative Counties of England and Wales'.

18   R.Y. Keers, *Pulmonary Tuberculosis: a Journey down the Centuries* (London, 1978), p. 73.

19   *Chief Medical Officer of Health Annual Report for 1919* (London, HMSO, 1920), p. 53.

20   For a history of the Association, see Glynne R. Jones, 'The King Edward VII Welsh National Memorial Association 1912-1948', in John Cule (ed.), *Wales and Medicine* (British Society for the History of Medicine, London, 1975), pp. 30-41.

21   PRO MH75/26.

22   *Western Mail* and *South Wales News*, 30 Jan. 1937.

23   PRO MH75/26.

24   R. Merfyn Jones, *The North Wales Quarrymen*, pp. 34-41.

25   *Parliamentary Debates, House of Commons*, vol.194, 1377, 51, (22 April 1926).

26   Subsequent research estimated respirable slate dust to contain 13-32% respirable quartz, still large enough to irritate the lung. J. R. Glover et al., 'Slate Dust in North Wales', p. 152.

27   *Wade Report*, p. 29.

28   Alice Stewart, 'Tuberculosis in Industry', p. 682.

29   *Western Mail*, 16 June 1927.

30   PRO MH96/1122. Wade Report, unrevised, 1926, Appendix 3 (not printed in final report).

31   Ibid., interview 19 July 1926, Dr Bradley Hughes of the Penrhyn Quarry Hospital, Bethesda. Also see Caernarvonshire Joint Sanitary Authority Meeting, *Liverpool Post*, 5 July 1927.

32   Wade Report, unrevised, Appendix 3.

33   *The Slate Trade Gazette*, Aug. 1927. Also reported in The *Cambrian News* and *Welsh Farmers Gazette*, 26 Aug. 1927. Road dust would, however, have included slate dust in the area.

34   *The Slate Trade Gazette*, Aug. 1927.

35   PRO MH96/1122, Welsh Board of Health, Memo 32A, Health Advisory Committee, 'Report on the Inquiry into the Occurrence of Respiratory Disease among Slate Workers in the Gwyrfai district'.

36   PRO MH96/1122, Letter from Wade to E. L. Collis, Mansel

Talbot Professor of Preventive Medicine, University of Wales, 3 June 1929.

37   PRO MH96/1122, Letter to Wade from W. Dairel, Cefn Mably Tuberculosis Hospital near Cardiff, 18 June 1929.

38   J. E. Tomley, *The Cambrian News*, 10 Dec. 1931.

39   PRO MH96/1056, Report of the Medical Committee of Memorial Association, 10 Dec. 1931, Memo. by Director of Research on the Incidence of Tuberculosis in Wales.

40   PRO MH55/1222, Dr J. W. Morris, evidence before 1937-39 Inquiry.

41   *Chalke Report*, pp. 74, 80.

42   PRO MH55/1236, Statement by Gwyrfai Rural District Council, 16 March 1938, for 1937-39 Inquiry. Also Dr V. Emrys Jones, Tuberculosis Officer, Caernarvonshire, *Transactions of 24 N.A.P.T. Conferences*, (1938), p. 71.

43   PRO MH55/1211, Dr Lloyd Owen, evidence before 1937-39 Inquiry.

44   J. Roberts, Llanberis, *Liverpool Post and Mercury*, 5 Aug. 1933.

45   PRO MH55/1222, Evidence of Dr J. W. Morris before Inquiry, (also reported in *Cambrian News*, 3 Dec. 1937); PRO MH55/1250, Evidence of Caernarvonshire County Council, 10 March 1938, pp. 194-207.

46   PRO MH55/1222, Letter by Clement Davies to Home Office, 2 Sept. 1938.

47   Report of meeting of Tuberculosis Association, *British Journal of Tuberculosis*, vol. 33 (1939), pp. 219, 220.

48   T.W. Davies, 'Silicosis in Slate Quarry Miners', *Tubercle*, vol.20 (1939), pp. 543-5.

49   *The Welsh Slate Industry — Report by Committee appointed by Ministry of Works*, Chairman Sir F. Rees, (London, HMSO, 1947), p. 15. (Shed workers in open quarries were included in the Silicosis Scheme Feb. 1946, p. 15).

50   *Wade Report*, p. 29.

51   *The Times*, 19 June 1933.

52   PRO MH96/1056, *Report of the Medical Committee of Memorial Association*, 10 Dec. 1931, (Memo. by Director of Research on the incidence of Tuberculosis in Wales, pp. 317, 318). Emrys G. Bowen, *Journal of the Royal Anthropological Institute of Great Britain and Ireland* vol. 63 (1928), pp. 363-399.

53   Montgomeryshire Insurance Committee meeting, *County Times*, 24 Oct. 1931.

54 PRO MH75/15; MH96/1056. J.E. Tomley, *The Manchester Guardian*, 15 July 1933.

55 Dr A. C. Watkin, *19 Annual Conference N.A.P.T.* (1933), p. 145. Kenneth Morgan points out that language differences in Wales implied much deeper cultural divisions, *Rebirth of a Nation, Wales, 1880-1980* (Oxford, 1982), p. 88.

56 *Liverpool Post and Mercury*, 5 Aug. 1933.

57 R. C. Hutchinson, 'Tuberculosis in a Welsh Country Community', *Tubercle*, vol. 2 (May 1921), pp. 345-9.

58 *Chalke Report*, p. 56.

59 PRO MH55/1208, Memo. of Evidence of Caernarvon Insurance Committee for 1937-39 Inquiry, p. 11. (Notification of tuberculosis cases had been made compulsory in 1912).

60 Ibid.

61 PRO MH55/1211. *Caernarvonshire Joint Sanitary Authority, 29 Annual Report for 1936* (1937), G. Lewis Travis and E. Lloyd, Recommendations, no. 13.

62 *Chalke Report*, p. 32.

63 *24 N.A.P.T. Conference* (1938), p. 70.

64 PRO MH75/15. *The Times*, 19 June 1933.

65 *Cambrian News*, 19 March 1937.

66 PRO MH75/26. J. Rowland Memo. for Welsh Board of Health, 8 March 1937 (waiting list in Wales on 1 July 1935 was 281 although it was pointed out (MH75/6) that patients were not often recommended for treatment owing to the shortage of accommodation).

67 For example, J.B. McDougall and N.D. Bardswell pointed out that of the 3,000 patients treated in sanatoria for tuberculosis under the London County Council scheme in 1927, 2,280 or 76% were dead by 1932. *Tubercle*, vol.17 (March 1935), p. 267.

68 See for example, *Tubercle*, vol. 11 (Sept. 1930), p. 540; vol. 16, (Oct. 1934), p. 31; vol. 18 (Aug. 1937) p. 522; vol.19 (Jan. 1938), p. 166.

69 M. Greenwood, 'The Epidemiology of Tuberculosis', *On the State of Public Health, Report of the Chief Medical Officer of the Ministry of Health for 1919* (London, HMSO, 1920), Appendix 5, p. 337

70 Wade Report, unrevised, Appendix 3.

71 Ibid.

72 *Chalke Report*, p. 18.

73 *Western Mail*, 19 July 1933.

74   Ibid.

75   *Liverpool Post and Mercury*, 13 Nov. 1934.

76   Ibid.

77   PRO MH55/1243.

78   Ibid.

79   Ibid.

80   *Clement Davies Report*, p. 36.

81   *Chalke Report*, p. 44.

82   R. Merfyn Jones, *The North Wales Quarrymen*, p. 31.

83   *Chalke Report*, pp. 17, 18, 43. *The Times*, 19 June 1933.

84   PRO MH55/1208, Evidence before 1937-39 Inquiry.

85   Ibid. See also V. Emrys Jones, *24 N.A.P.T. Conference*, (1938), p. 71; and Dr Lloyd Owen, *Caernarvon and Denbigh Herald*, 4 Dec. 1931.

86   *13 N.A.P.T. Conference*, (1927), p. 16; *14 N.A.P.T. Conference*, (1928), p. 144.

87   Synopsis of films, *Annual Report N.A.P.T. Council for 1924* (1925), Appendix 2, p. 29.

88   PRO MH55/1222, Letter from E. Andrewes, 28 March 1938, to Clement Davies.

89   PRO MH55/1222, J.W. Morris, evidence before 1937-39 Inquiry, p. 5.

90   PRO MH55/1211.

91   *The Cambrian News*, 2 May 1924; *Clement Davies Report* , p. 93; F.R.G. Heaf, *Symposium of Tuberculosis*, p. 711.

92   PRO MH96/1056.

93   PRO MH55/1222. Letter from Hoare to Clement Davies, 14 Nov. 1938.

94   PRO MH55/1222. Evidence of Morris.

95   *Rees Report on Welsh Slate Industry*, p. 15; Jean Lindsay, *North Wales Slate Industry*, p. 298.

96   Wade Report, unrevised, Appendix 3.

97   P. Wright and A. Treacher (eds.), *The Problem of Medical Knowledge, Examining the Social Construction of Medicine*, (Edinburgh, 1982), Introduction by Wright and Treacher, p. 4.

98   *Chalke Report*, p. 32.

1 Strike committee of the Matchmakers' Union, 1888. During the 1880s there was increasing concern with the chronic inflammation of the jaw, known as phossy jaw. In 1906 European countries banned the manufacture and import of white phosphorous matches. From A. Besant, *An Autobiography* (London, 1893), p. 336. Reproduced by permission of the Wellcome Institute library, London.

2 Fork grinding at Sheffield, 1860. From *Illustrated London News*, 8 March 1866. Reproduced by permission of the Wellcome Institute library, London.

3   Sir Thomas Legge (1863-1930). The first medical inspector of
factories. Reproduced by permission of the Wellcome Institute
library, London.

4   Ludwig Teleky (1872-1957). The first Prussian medical inspector of factories. From the papers of Anna Teleky, in the Archive of Industrial Pathology, University of Bremen.

5   Working rock drills. Wheal Agar Mine, Cornwall 1898. Reproduced from the Burrows Collection, Royal Institution of Cornwall, Truro.

6   Hand stemming of TNT into a shell, causing exposure of the
worker's hands and soiling of the floor. Reproduced from 'TNT
Poisoning and the Fate of TNT in the Animal Body', *Medical
Research Council, Special Report Series no. 58* (London, HMSO,
1921).

7   Chilwell stemming machine in operation. Reproduced from
     'TNT Poisoning and the Fate of TNT in the Animal Body',
     Medical Research Council, Special Report Series no. 58 (London,
     HMSO, 1921).

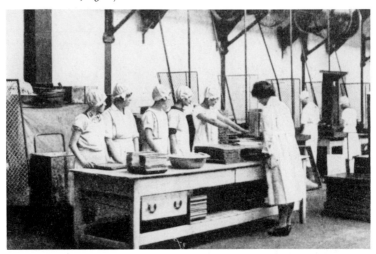

8   Guarding against 'Baker's itch'. A doctor examining the hands
     of workers in a biscuit factory for signs of eczema produced by
     long contact with some of the ingredients. Reproduced by
     permission of the Wellcome Institute library, London.

9　First aid in the work-room. Reproduced by permission of the Wellcome Institute library, London.

Last Year's Accidents
97,143
Equivalent to:
1 PERSON INJURED EVERY 1½ MINUTES
(4 4 HOUR WEEK)
843 Fatals
Equivalent to:
1 PERSON KILLED Every 2 hrs. 43 Minutes
BAD ENOUGH, BUT LOOK AT THIS
ACCIDENTS WHICH COULD 80% EVEN
HAVE BEEN AVOIDED 20%
TOTAL ACCIDENTS
IS ACCIDENT PREVENTION WORTH WHILE TO YOU ?

Tons of Money
£12,000,000 LOST
by Industrial Workers in Wages
THROUGH ACCIDENTS IN 1921
If **YOU** incurred an accident some of it was **YOURS !**
We don't want carelessness to hand in a similar bill this year !!

DO IT THE SAFE WAY
An efficient worker is always a safe worker. Have YOU gained the reputation of being a ·:· safe worker ? ·:·
RECKLESSNESS IS NO INDICATION OF COURAGE

WEAR YOUR GOGGLES
YOU CAN EAT WITH FALSE TEETH WORK WITH FALSE ARMS WALK WITH FALSE LEGS
BUT YOU CAN'T SEE WITH FALSE EYES
PROTECT YOUR EYES

CAN ACCIDENTS BE PREVENTED
This Record Should Convince You
IF 4 Months Why Not 12 ?
ONE FIRM WELL ON THE WAY!

A CAT HAS NINE LIVES
YOU *have only One Take good care of it.*

10　Accident prevention posters. Reproduced by permission of the Wellcome Institute library, London.

11   Model ambulance and recovery rooms, Home Office Industrial Museum. Reproduced by permission of the Wellcome Institute library, London.

12   Clinica del Lavoro 'Luigi Devoto', Milan. Reproduced by permission of the Wellcome Institute library, London.

7 A PATIENT IN NEED OF CARE: GERMAN OCCUPATIONAL HEALTH STATISTICS

Rainer Müller

The medical statistics of the Federal Republic of Germany are underdeveloped in comparison with other industrial countries.[1] Occupational mortality and morbidity statistics do not cover all employees. The annual statistics of the sickness insurances *(Gesetzliche Krankenversicherungen)*, even those for absenteeism, are not arranged according to occupation. The publications of pension funds *(Rentenversicherungen)* on invalidism are not categorised by occupation or employment sectors. Even the accident insurances *(Unfallversicherungen/Berufsgenossenschaften)* have only recently begun to publish statistics of accidents at work and occupational diseases according to occupation.

A few researchers in social medicine have, on their own initiative, recently begun to compile statistics according to categories such as occupation, and economic sectors.[2] They have been using records of sickness and pension funds. Although in 1980 200,000 million marks were spent on the health sector, only a minute fraction of this sum is available for public health and epidemiological research. Demographic problems, and difficulties in financing health services have prompted the Federal Government to fund two research programmes, which provide health statistics comparable with Scandinavia, Britain and the U.S.A. Yet, no legislation has been forthcoming. The demographic problems, especially those associated with health risks, are shown in statistics of the labour market which are projected to 1990. Between 1978 and 1990 2.8 million employed males will retire; 33% will die prematurely before they reach normal retirement age; and 37% will become prematurely invalided due to chronic disease. Only 30% will draw normal pensions.[3]

Centenary celebrations of the social insurance laws of 1881 to 1884 have not explained why statistics of occupational health have not been developed and this is the issue which I should like to consider here. Although there are some historical studies of the development of medical statistics, there is no social history of German occupational health statistics. An historical survey of attempts to report on industrial pathology should include reference to the efforts of the labour movement. In 1905 a journal edited by Max Weber judged that the social and medical statistics of the workers' associations and unions were the most important and only reliable source of information on work and health conditions.[4]

For the history of investigations into social inequalities in death and diseases, the origin of population statistics is pertinent. The very term *Statistik* expressed the state's interest in data on its population. Medical statistics developed in the context of population statistics, which began in the late seventeenth and early eighteenth centuries. In Prussia, in 1685, a central administrative department was established to supervise all aspects of medicine *(Arztney-Wesen)*. [5] The first census was taken in 1719, but it was not until 1748 that one was annually conducted. The first census of occupations in Prussia was made in 1722. In 1741 Süssmilch published an influential book on population and medical statistics.[6] Since 1764 the church has been required to keep statistics of births, marriages and deaths. The Prussian Statistical Office was founded in 1805 by Minister Stein. According to his instructions there was to be reporting of population movements, hospitals and other aspects of health, including working conditions. Official statistics were only published from 1839. From 1859 an annual statistical report, the *Preussische Statistik*, was published. The Journal of the Prussian Statistical Office was edited from 1861 by its Director, Ernst Engel.

Statistics of occupational mortality need two preconditions: a census by occupation, and statistics of mortality which include details of age, sex and occupation. (These have been carried out in England and Wales since 1851). In Germany it was impossible to co-ordinate and link these two requirements. In 1877 the directors of the various statistical offices concluded that existing occupational statistics were totally inadequate.[7] The Reich Statistical Office complained in 1878 of the difficulties of compiling a uniform classification of occupations for a census. The paucity of German social statistics was also complained of by Karl Marx in *Das Kapital* in 1867. He accused the German states of deliberately concealing the extent of poor social conditions.[8]   A census of occupations and industries in all German states using uniform methods and categories was first carried out in 1895. From 1899 in Württemberg the mortality statistics were linked to occupation. For Prussia such statistics, for a limited number of occupations without cause of death, exist for 1906 to 1908.[9] The first occupational mortality statistics were compiled in 1955 in the Federal Republic and were based on the 1950 census. This study was published in 1963 and has remained the only one of its kind.[10]

One important reason why routine collection of occupational health statistics has not occurred has been the conflicting interests of the federal states. After the founding of the *Reich* in 1871, it was not possible to establish a central organ directing all health policy, and even today there is no central Ministry of Health in Bonn. It was not even possible to create compulsory and uniform registration of procedures for the certification of causes of death.

There was no lack of demand for a central Reich Ministry of Health. The medical reform movement associated with Virchow in the 1840s demanded a central ministry from 1848. In 1871, a memorandum, commissioned by Bismarck, concluded that a central ministry of health was not practicable but recommended a central office for health statistics. This had been demanded by German medical statistical societies in the 1860s. A commission to prepare medical statistics for the German Empire was established in 1874 and reported in 1875. It declared that morbidity statistics for the whole German population was impossible. It suggested, however, that statistics of epidemics, infectious diseases, of hospital patients, army recruits and certain occupations covered by special insurances should be gathered. Mortality statistics for the whole population according to cause of death, age and sex were to be compiled.

A classification of 80 diseases was suggested. The draft of a Bill to determine causes of death proposed the inclusion of occupation as a category. General occupational mortality statistics were not considered necessary. Occupational mortality statistics were only recommended in connection with the morbidity statistics of particular occupations for which special insurances were in operation: such as miners, postmen, railway officials and other workers who were members of, for instance Friendly Societies *(Freie Hilfskassen)* or private insurance schemes. A central statistical office was to analyse and publish results. For doctors, compulsory registration of infectious and contagious diseases was proposed. However, doctors protested that these recommendations violated their professional freedom. The Commission's proposals were sent to Bismarck to be made law, but they were not passed. When the Imperial Health Office was founded it intended to improve statistics.[11] Taking on responsibility for statistics in 1876, the Health Office's sole function was to advise the central government. Bismarck always wanted the Office to exert a strong centralising influence. However, instead of selecting a qualified expert in public health or medical statistics like Pettenkofer as Director, the government appointed a military doctor, Heinrich Struck, who was also Bismarck's doctor. The Office initially considered medical statistics to be of prime importance, but this later gave way to bacteriology.

In 1875 those heads of friendly societies in Berlin who were socialists, called for a Central Office of Statistics and demanded occupationally relevant health statistics. They combined the call for occupational statistics with meetings of doctors employed by friendly societies in large cities to discuss the implications of statistics for prevention and therapy.[12] Even today these demands have not been realised. The Imperial Health Office failed during the *Kaiserreich* and Weimar Republic to produce valid social medical statistics. The present Health Office has also so far failed to develop them. In 1854 a Factory Inspectorate was instituted by the Prussian state, to monitor health risks in factories and to supervise the 1839 child labour and factory legislation. In 1872 the Prussian administration realised that the inspectorate was ineffective and insufficient. As a result of the growing labour movement and strikes in 1873 the Prussian bureaucracy conducted a general revision of policies for the protection of labour. It intended to provide comprehensive information on working conditions and health risks. In 1872 the Prussian Ministry of Trade attempted to obtain statistics on fatal and non-fatal accidents at work. The Ministry also wanted accurate information from sick funds about diseases which might be caused and exacerbated by working conditions. Resenting this interference, factory sick funds and factory doctors boycotted the survey of work-related diseases, and the employers refused to compile the accident statistics. The Bill of 1876 to reform the Imperial Trade Regulations *(Reichsgewerbeordnung)* provided for systematic registration of accidents at work and work-related diseases, but employers' opposition blocked this initiative.

Although factory inspection was extended from 1891 there were still too few inspectors. Their role was largely advisory, and they dealt with machinery accidents rather than occupational diseases. Except in Baden the inspectorate's reports remained unsatisfactory and vague. Even today there is no continuous and systematic reporting of the nature and extent of health risks at work. Occupational health and statistical studies were occasionally developed for specific industries when doctors were appointed to the factory inspectorate. The first medical factory inspectors were Holtzmann in Baden in 1906, Koelsch in Bavaria in 1909, and Teleky in Prussia in 1921. The extension of the factory inspectorate at the beginning of the 1890s was linked to initiatives to produce comprehensive social medical statistics.

In 1897 a *Reichskommission für Arbeiterstatistik* (Commission for

Statistics of Living and Working Conditions) was set up with only one socialist member, Schippel, who was later replaced by Molkenbuhr.[13] The chairman, Woerishofer, had a good reputation as Director of the Factory Inspectorate in Baden. He wished not only for the preparation of statistics but also for commissions and hearings, citing British examples as precedents. He also wanted an institution independent of the Imperial Statistical Office *(Kaiserliches Statistisches Amt)* with a decentralised administration, but he failed to achieve this. The *Reichskommission für Arbeiterstatistik* first investigated conditions in bakeries. This was a reaction to the S.P.D.'s inquiry into baking and health conditions and was published in 1890. The *Kommission für Arbeiterstatistik* was opposed by employers and controversies arose even within the S.P.D. despite the support of its leader, August Bebel. Relatively early in the second half of the nineteenth century miners' insurance funds *(Knappschaftskassen)* produced reports of a higher quality than other funds. It is appropriate to consider in greater detail the achievements of the sickness insurances *(Gesetzliche Krankenversicherungen)*. Bebel welcomed methods of enquiry based on the English model and asked for an independent state office of labour statistics. The scope of the *Kommission für Arbeiterstatistik* was continually restricted. In 1902 a special department was established within the Imperial Statistical Office and the *Kommission für Arbeiterstatistik* became a subsidiary committee. Until 1914 some surveys were carried out on working hours, but no competent institution could be set up and consequently the statistics have remained inadequate.[14]

There have been complaints that the pension funds *(Rentenversicherungen)* have published statistics which have been grossly inadequate. Furthermore, the statistics of the accident insurance funds *(Unfallversicherungen/Berufsgenossenschaften)* on accidents at work and occupational diseases do not give sufficient information on risk at work. Before 1925 there was no compensation for occupational diseases. At first, there were only 11 occupational diseases; now 55 are on the list. There has never been reporting of unspecific work-related diseases.

By 1850 the first studies based on records of sickness insurances in Germany appeared as reports on hospital treatments. They attempted to investigate the influence of occupation on the health of artisans and craftsmen.[15] The first worthwhile attempt to compile occupational medical statistics based on insurance records was carried out in the context of the workers' self-help movement. From 1849 until 1853 the

Health Care Association *(Gesundheitspflegeverein)* of the Berlin workers' union *(Arbeiterverbrüderung)* was a co-operative organisation of workers and doctors.[16] Unfortunately, no other such associations have been formed. The organisational structure and the activities of the association should be a model for a democratic health system, emphasising the social conditions of health. This self-help organisation was created for sickness benefit and for medical care for 10,000 members. The Health Care Association was run co-operatively by a medical committee led by the radical doctor Salomon Neumann and the leader of the Workers' Union. The Association was fundamentally different from that envisaged by the doctors of the medical reform movement. The liberal medical reformer, Virchow, demanded state medical care. In contrast, the Workers' Union doctors demanded political emancipation and autonomous organisation. Within the Health Association under Neumann the basis for occupational health statistics was established, and he compiled comprehensive statistical reports. Neumann hoped that the statistics would illustrate the general influence of living conditions and the impact of specific occupations on morbidity and mortality. The King of Prussia banned the Association in 1853 because he disliked its 'dangerous socialist and communist activity'.

Neumann continued to be active in the field of medical statistics. At the International Statistical Congress in Berlin in 1863, he described a comprehensive plan for the organisation of a census and compilation of social and medical statistics. In 1856 he analysed the records of 40,000 members of 67 sickness insurance funds in Berlin. A law of 1876 attempted to standardise and place under state control all sickness funds. The law made provision for regular reports with uniform criteria of morbidity and mortality to be sent to a central office, but these statistical reports were never made. There were only statistics of the number of insurances and the size of their membership. Medical statisticians had great hopes for this law. This is evident from a memorandum by the Director of the Prussian Statistical Office, Ernst Engel, for the Ninth International Statistical Congress in Budapest in 1876 entitled 'The statistics of morbidity, invalidity and mortality and the statistics of accident insurances and pension funds of the employed'.[17] The examples of Neumann and Engel show that in the years preceding the social legislation of the 1880s in Germany, there were experts competent to produce comprehensive social and medical statistics.

In 1883 the Sickness Insurance Act, which is still in force today, was passed. The law did not require compilation of comprehensive medical statistics. The codification of social insurance statutes of 1911 (the *Reichsversicherungsordnung*), also missed the chance to legislate for such statistics. Certain civil servants attempted to collect these. In 1902 a memorandum on occupational morbidity statistics for the Reich administration recognised the urgent need for such statistics. The organisation and expertise were certainly available. The Imperial Statistical Office in association with the Imperial Health Office compiled a comprehensive study on morbidity and mortality in Leipzig from 1886 to 1905. Records of 900,000 male and 200,000 female members of the Leipzig municipal sick funds were analysed using 180 male and 79 female occupational categories. The data were classified under 355 diseases, which included accidents at work.[18]

Of special value was the study by the Director of the statistical office of Frankfurt am Main using the records of the Frankfurt and Bockenheim sick funds for the years 1895 and 1896.[19] Whereas the Leipzig study reported on the sick who were absent from work, the Frankfurt study reported on the sick who were ill but continued to work. Until 1933 wherever health matters relating to work arose, these studies were regarded as standard. In the Federal Republic in the late 1970s, in connection with our research in Bremen and Berlin, we tried to match the high quality of these model experiments. Today, we are in a position to implement the dream of medical statisticians in the twenties of carrying out prospective longitudinal studies of cohorts based on records of sickness insurance funds.

In the 1920s Ludwig Teleky, the Prussian medical factory inspector in Düsseldorf worked to improve insurance-based health statistics.[20] At every level, scientific, political and administrative, he urged the development of such statistics. In his writings he showed that insurance records gave information about groups of employees at risk. It seemed possible to him, on the basis of this data, to identify typical career and disease profiles. In 1933 this work, linking the sick funds and occupational hygiene, ceased. Because Teleky was Jewish and a socialist he was dismissed; he fled to Vienna and, in 1939, to New York. In the early years of the Federal Republic, the scientific and social political traditions of social medicine, associated with the names of Virchow, Neumann, Zadek and Teleky were not revived. Nazism had perverted social medicine.

My conclusion is that efforts to improve occupational health statistics have been unsuccessful. Bismarck's social legislation of the 1870s and 1880s was accompanied by an attempt by the state administration to produce occupational health statistics while influential associations pressed for reform of health statistics. The reasons for this failure are complex. These include resistance of employers and localism of the many German states. The shift in public health to bacteriology in the 1880s obstructed the demand for medical and social statistics. Health was only a secondary priority in trade union policy which, after 1900, placed too great a reliance on the state and on professional experts. Safety at work legislation was primarily technical. Occupational medicine was incorporated into the factory inspectorate relatively late and could not gain a foothold in medical education. Occupational medicine was oriented too much towards the natural sciences. The takeover of health policy by doctors' professional organisations was a result of the operation of social insurance mechanisms and the doctors' concentration on the individual patient. Collective experience of health risks at work was translated into individual problems and specific instances. A social aetiology of disease could not be developed against the dominance of clinical medicine. These reasons must be considered in the context of general political developments in Germany: the imperialism of the *Kaiserreich,* two world wars, economic crises during the Weimar Republic, fascism, and the conservatism of the post-war Federal Republic.

Although I am complaining about the lack of social and medical statistics in Germany I do not regard statistics as self-evident and unbiased, when used to demonstrate social inequality. It cannot be assumed that the state employs social statistics to evaluate its own policies. It should be borne in mind that official statistics are produced and based on methods and categories within particular ideological contexts. Social and medical official statistics are necessary for demystification. Without such state-gathered statistics a Black Report on inequalities in health in Britain would not be possible![21]

Any report on industrial pathology that claims to give a true picture must have, in addition to official statistics, contributions from critical epidemiology, as well as an input from the labour movement. Attempts must be made through political lobbying to ensure that official statistics reveal conditions realistically. In the FRG during recent years the labour movement has begun to compile its own reports on working conditions and health hazards. The *I.G. Metall*, the steel workers' union, has carried out extensive investigations.[22] The results have been used in labour disputes in factories to obtain improvements in working conditions, and to mobilise workers, and have contributed to the development of social and health policies in general.

I wish to thank Paul Weindling and Kate Williams for helping me to express my views in acceptable English.

Notes

1   Bundesminister für Jugend, Familie und Gesundheit (ed.), *Daten des Gesundheitswesens, Ausgabe 1980* (Stuttgart, Berlin, Köln, Mainz, 1980).

2   F. Eggeling, *Zur Epidemiologie der Berufskrankheiten* (Dortmund, 1980); M. Blohmke and K. Reimer, *Krankheit und Beruf* (Heidelberg, 1980); R. Müller, E. Bergmann, A. Musgrave and K. Preiser, *Berufliche, wirtschaftszweig- und tätigkeitsspezifische Verschleissschwerpunkte. Analyse von Arbeitsunfähigkeitsdaten einer Ortskrankenkasse* (Bonn, 1981); L.v. Ferber and W. Slesina, 'Arbeitsbedingte Krankheiten' in *Sozialpolitik und Produktionsprozess* (Köln, 1981), pp. 37-61; A. Georg, R. Stuppardt and E. Zoike, *Krankheit und arbeitsbedingte Belastungen* (2 vols., Essen, 1981-1982); R. Müller, F. Schwarz, H. Weisbrod and P. König, *Fehlzeiten und Diagnosen der Arbeitsunfähigkeitsfälle von neun Berufen* (Dortmund, 1983).

3   H. Bloss, 'Abgänge sozialversicherungspflichtig beschäftigter Arbeitnehmer nach Berufen bis 1985 und 1990', *Mitteilungen des Instituts für Arbeitsmarkt- und Berufsforschung* (1979), pp. 166-177.

4   D. Tutzke, 'Die Rolle der medizinischen Statistik in der Sozialhygiene in Deutschland vor 1933', *Zeitschrift für ärztliche Fortbildung*, vol. 66 (1972), pp. 1158-1161; D. Tutzke, 'Zur geschichtlichen Entwicklung der medizinischen Statistik', *NTM*, vol. 11 (1974), pp. 72-79.

5   A. Fischer, *Geschichte des deutschen Gesundheitswesens* (2 vols., Berlin, 1933).

6   J. P. Süssmilch, *Die göttliche Ordnung in den Veränderungen des menschlichen Geschlechts, aus der Geburt, Tod und Fortpflanzung desselben erwiesen* (Berlin, 1741).

7   Fischer, *Geschichte des deutschen Gesundheitswesens*, vol. 2, p. 474.

8   K. Marx, *Das Kapital, Vorwort zur ersten Auflage* (Hamburg, 1867).

9   F. Prinzing, *Handbuch der medizinischen Statistik*, 1st edn (Jena, 1906), p. 475, 2nd edn (Jena, 1931), pp. 611, 616.

10 Statistisches Bundesamt, *Beruf und Todesursache (Ergebnis einer*

Sonderzählung 1955), *Fachserie A, Bevölkerung und Kultur, Reihe 7, Gesundheitswesen* (Stuttgart, Mainz, 1963).

11   M. Stürzbecher, '100 Jahre Forschen für die Gesundheit. Vom Kaiserlichen Gesundheitsamt zum Bundesgesundheitsamt', *Die Berliner Ärztekammer*, vol.4 (1976), pp. 147-154.

12   *Neuer Social-Demokrat*, no. 137 (19.11.1875), no. 138 (21.11.1875), no.139 (24.11.1875).

13   W. Bocks, *Die Badische Fabrikinspektion. Arbeiterschutz, Arbeiterverhältnisse und Arbeiterbewegung in Baden 1879 bis 1914* (Freiburg, München, 1978).

14   P. Mombert, 'Neuere sozialstatistische Erhebungen deutscher Arbeiterverbände', *Archiv für Sozialwissenschaft und Sozialpolitik*, vol. 21 (1905), pp. 248-265.

15   K.-H. Karbe, 'Die Entwicklung der Arbeitsmedizin in Deutschland von 1780 bis 1850 im Spiegel der zeitgenössischen medizinischen Literatur', Dissertation der Karl-Marx-Universität, Leipzig, 1978.

16   E. Hansen, M. Heisig, S. Leibfried and F. Tennstedt, in cooperation with P. Klein, L. Machtan, D. Milles and R. Müller, *Seit über einem Jahrhundert...: Verschüttete Alternativen in der Sozialpolitik* (Köln, 1981).

17   E. Engel, 'Die Statistik der Morbidität, Invalidität und Mortalität, sowie der Unfall- und Invaliditätsversicherung der Erwerbsthätigen', *Zeitschrift des Königlich Preussischen Statistischen Bureaus* (1876), pp. 129-188.

18   Kaiserliches Statistisches Amt, *Krankheits- und Sterblichkeitsverhältnisse in der Ortskrankenkasse für Leipzig und Umgebung* (4 vols., Berlin, 1910).

19   H. Bleicher, 'Frankfurter Krankheitstafeln. Untersuchungen über Erkrankungsgefahr und Erkrankungshäufigkeit nach Alter, Geschlecht, Zivilstand und Beruf auf Grund des Materials der Ortskrankenkassen zu Frankfurt a.M.', *Beiträge zur Statistik der Stadt Frankfurt am Main*, neue Folge, no. 4 (1900).

20   L. Teleky, 'Die Krankheitsstatistik der nach dem 'Rheinischen Schema' arbeitenden Krankenkassen 1922-1931', *Archiv für Gewerbepathologie*, no.5 (1934), pp. 764-809.

21   P. Townsend and N. Davidson (eds.), *Inequalities in Health. The Black Report* (Harmondsworth, 1982).

22   Industriegewerkschaft Metall, Bezirksleitung Stuttgart, *Werktage müssen menschlicher werden* (Stuttgart, 1979).

8 'CORONARY' HEART DISEASE - A DISEASE OF AFFLUENCE OR A DISEASE OF INDUSTRY?

Mel Bartley

> ... existing conditions of work ... impose strains which, when endured too long, are beyond physiological tolerance ... conditions thus call for amendment ... In the field of personal hygiene, the organisation of holidays, leisure, exercise and pleasurable relaxation is as sensible as attention to sanitary habits and balanced dietaries ... Healthy living should be promoted through a more precise physiological and psychological understanding of man and of his individual and social needs. [1]

> Surely a prudent diet, with reduction of fat energy and control of obesity and sloth are sound policies for improving the public health. [2]

These two extracts, written over thirty years apart, both refer to preventive strategies directed at the same disease, 'coronary' heart disease. They represent the prescriptive implications of two different theories about what sort of disease this is and how it is caused. These theories, in turn, rested upon two distinct pictures of the kind of person who 'gets' the disease. The idea that different kinds of people contracted heart disease led to a reformulation of medical ideas about its nature, not just its aetiology, but the very pathology involved.

A re-reading of the routinely collected mortality statistics and the literature on the problems of death certification throws considerable doubt on the idea of *one*, clearly distinguishable 'modern epidemic of heart disease'. On the contrary, chronic disease of some cardio-respiratory nature appears to have:
a) been the leading cause of death in middle age for men and women since 1847,
b) been expressed in vital statistics which were sorted and re-sorted over the period 1847-1970 by changes in the very aetiological theories that the statistics themselves were used to 'prove',

c) accounted for more of the overall fall in the official general mortality rate for middle aged men than is usually acknowledged in modern community medicine literature — at some stages heart disease declined as fast as the infectious diseases.

These points can be illustrated by reference to official reports of the Ministry of Health and the Registrar General. In his Report for the year 1921, the Chief Medical Officer to the Ministry of Health was unhappy about the health of middle aged persons:

> Seventy years ago, wrote Sir George Newman, men
> of 45 had on average 22.76 more years to live. There
> was no improvement on this figure for thirty years.
> On the contrary, it seemed to decrease. [3]

A Swede, according to the league tables of national health achievements presented in this and other reports of the Chief Medical Officer (CMO), could expect to live three years beyond his forty-fifth birthday more than a Briton. Newman suggests:

> We must scrutinise very closely the causes of
> mortality in later maturity ... In England and Wales
> in the decennium 1901-1910 the largest single cause
> was pulmonary tuberculosis ... (next) non-
> tubercular respiratory diseases. [4]

Also, while he rejected the view of Sir John Simon that poverty and industrial hazards could have anything to do with this, he noted that mortality in this age group was unevenly spread across the country. Death rates were higher in industrial areas of the North West such as Lancashire. Nor can the climate be to blame, because:

> The very high mortality from respiratory disease of
> the northern manufacturing towns is a serious
> feature of our vital statistics, and not wholly due to
> climate, as a comparison of Lancashire and
> Cumberland proves. [5]

In the Report for the year 1925, the Registrar General points out that: 'Bronchitis is particularly fatal to the urban population in the North of England', and that the cold, which is presumed to affect respiratory disease, does not even compensate the unfortunate northerners by protecting them from an equally high mortality from heat-associated epidemics such as infantile diarrhoea. Laconically, Newman admits that, 'Simultaneous excesses of 74 per cent for mortality supposed to be promoted by cold and over 100 per cent for mortality known to be promoted by heat cannot both be explained on the score of climate'. [6]

In the Report of the Registrar General for England and Wales for the year 1925 we find no commentary on heart disease at all. It seems to pass as an unremarked fact that, according to the tables at the end of the Report, something called 'organic heart disease' has been the highest proportional cause of death in the adult population for some time. In these tables, 'organic heart disease' is listed as causing 102 per thousand deaths, more than any other single category. In the Registrar General's Report for 1926, a new disease entity begins to emerge. In 1921, a new subheading of the International Classifications of Diseases (ICD) had been created: number 91 (b), 'arteriosclerosis cerebral vascular lesion'. It now became officially possible for someone to die of 'arteriosclerosis' or hardening of the arteries, whereas before, this cause had always been listed as 'stroke' (cerebrovascular lesion) without further elaboration. The Registrar himself comments that the younger, more recently trained and up to date doctors would tend to prefer the 'arteriosclerosis' term to the old fashioned 'stroke' when filling in a death certificate. The return of the younger doctors from the field of battle after 1918 was held to be the cause of a sudden apparent rise in the number of deaths from 'arteriosclerosis'.

There was clearly a debate on the 'rise' in the number of deaths apparently associated with degenerative changes (such as 'hardening') to the blood vessels, and the Registrar is obliged to address it, despite his own caveats on diagnostic fashions:

> Overeating is recognised as playing an important role in the causation both of arteriosclerosis in general and of cerebral haemorrhage ... the question whether, and to what extent, mortality from cerebral haemorrhage is increasing is of importance, because this form of mortality is well fitted to serve as an index of that from degenerative circulatory diseases in general. From time to time, cries of alarm are raised at the increase of recorded mortality from degenerative diseases of the circulatory system. However, 'arteriosclerosis' is rapidly replacing 'old age' in certification . . . And similarly, 'myocardial degeneration' is much oftener mentioned now than in former years on the death certificate of sufferers from chronic bronchitis, etc. As one of the current rules of classification prefers heart to respiratory disease if the two appear on the same death certificate, this change in vogue of certification

results in the transfer of all deaths from bronchitis etc
to heart disease. [7]

This Report, for the year 1926, is the first one which devotes a special section of its commentary to heart disease, though the Registrar General acknowledges it to be 'as usual, larger than any other item on the list of causes'.

Further research will be necessary in order to discover with whom the Registrar General was engaging in this debate over an epidemic of heart disease between 1911 and 1925. But he is insistant that it has two main causes: firstly, the ageing of the population, and secondly, the growing unrespectability of leaving 'old age' as a diagnosis on a death certificate (and perhaps also 'bronchitis'). 'Therefore' he concludes, 'alarmist pronouncements as to increase of mortality from heart disease by the 'stress and worry of modern life' may be met with the observation that it is declining.' He proves this point by replacing crude with age-standardised death rates from heart disease, which results in a picture more congruent with his general argument.

As the economic recession of the 1920s deepened, a deterioration in the health of middle aged persons became less easy to ignore. In his Report for 1927, the Registrar General admits that the death rate from heart disease has risen in some real sense, though he attributes this to an epidemic of influenza. He re-emphasises the effect of the 1921 coding rule changes which require that cardiac causes be regarded as more fundamental than respiratory causes when the two are present. It seems that the school of thought which held the rise in 'heart disease' to be a result of stressful social conditions is still powerful enough to require an answer in the official reports. By 1927, however, rises are visible in the official statistics for all major causes: rheumatic as well as 'degenerative' heart disease, bronchitis, pneumonia, tuberculosis, ulcers.

The Registrar General's Report for the year 1930 does not admit defeat. It states:

> the recent increase of crude mortality from heart
> diseases is due, amongst other causes, to the
> increasing age of the population, and to rapid
> increase in the record of myocardial degeneration in
> certification of the death of old people ... When
> allowance is made for the disturbing influences
> mentioned ... the increase ... is reduced ... [8]

One of the disturbing influences was another change in the rules for coding death certificates, made in 1929. From then on, death certificates stating 'myocardial degeneration with arteriosclerosis' were to be coded to heart disease, rather than to diseases of the blood vessels generally, as had been the practice. The heart thus became a focus for a disease process which had been observed and speculated upon for some time, but formerly diffused throughout the body. With this coding change, coupled with the 1921 stipulation about bronchitis, two very common diseases now became located in the heart, both the former 'bronchitis' and the former 'arteriosclerosis'.

The major step that remained to be taken before the modern pathological picture of coronary heart disease was accomplished, was to see the changes or degeneration of the heart muscle itself as 'due to' atherosclerotic occlusion of the coronary arteries. But in the 1930s, the majority of 'heart disease' was still classified as 'degenerative' rather than 'anginal' (i.e. due to coronary artery blockage). Notwithstanding this, by the 1930s the conceptual framework had been established for the delineation of a major cause of death associated with degeneration and failure of the heart 'due to' sclerotic and fatty changes in the arteries supplying blood to the heart muscle. In this way, the muscular failure of the heart, which is the end point of many diseases, can be attributed mainly or solely to the state of the constituents of the blood or blood vessel walls. It follows from this that theories about diet, physical activity and other aspects of lifestyle can be used to account for excessive cholesterol, insufficient thrombolytic factor, unbalanced High Density Lipoprotein to Low Density Lipoprotein ratio, absence of compensating collateral vascularisation and so forth, according to which version of the theory you espouse. The heart will then be expected to bear the 'stress of modern life' as long as the personal health practices of individuals have kept their coronary arteries in a healthy enough state. Thus the picture is almost complete, of a 'modern' disease, which, beginning at the time of the economic depression of the 1920s and 1930s, is about to become the commonest cause of death in men of working age, a picture of a disease which allows it to be attributed to rising standards of living.

This situation was reflected in the Reports of the Chief Medical Officer for the year 1932 and 1933. Newman wrote in his Report for 1932:

> The great constitutional diseases, which bear in particular a social and domestic significance, have declined proportionately within the last ten years. In 1923, tuberculosis was the cause of 92 in every 1000 deaths, in 1932 it caused 69 in every 1000 deaths . . .

> deaths from diseases of the nervous system fell from
> 107 deaths per 1000 in 1923 to 84 in 1932; and
> bronchitis and pneumonia, which caused 149 per
> 1000 in 1923 declined to 113 per 1000 in 1932. There
> has, of course, been an increase in the incidence or
> mortality of certain other diseases. But the morbid
> conditions and their mortality rates referred to in the
> Tables ... indicate that there is no evidence from
> mortality in support of exceptional physical
> deterioration in recent years owing to social
> circumstances. [9]

It should be noted here that the use of the word 'proportionately' and of proportional mortality rates (deaths from disease 'x' per 1000 deaths from all causes) rather than death rates per 1000 population is strategic. It avoids revealing the fact that per head of the population, deaths have risen. The CMO is forced into a position where he must argue for continued improvement in the health of the nation, not from the mortality experience of the nation, but by juggling around different causes of death which carry different (politically relevant) assumptions about aetiology. The 'certain diseases' which have increased are heart disease and cancer.

In his Report for 1933, Newman, (still the CMO) is still insisting that: 'the death rates which are most sensitive and susceptible of social or physical degeneration of a people remain steady and low'. [10] With the privilege of hindsight we can see, from Logan's tables [11] published in 1950, of mortality from 1848 to 1948, that the all-causes death rate for middle aged persons was rising in 1932-33, for the first time since 1900. But Newman blithely insists:

> It is said, with truth, that the number of deaths
> attributed to cancer is increasing, that a large
> proportion of deaths is due to heart disease, ... they
> are important but they do not provide evidence of
> national deterioration. [12]

This sounds uncannily similar to the tone adopted in reports on child health documented by Charles Webster in his analysis of the 'Healthy or Hungry Thirties?' [13] The ominous death rate cannot be absolutely denied, but instead it can be attributed to a disease of 'no social or domestic significance.' It is the argument of this paper that it was the establishment of 'heart disease' as a disease 'of no social significance', rather than any particular biological phenomenon, which accounts for its establishment as the modern epidemic which explained the failure

Table 1

Coronary, non-coronary, and total heart disease: rates per million men aged 45-64

Year	non-coronary heart disease	'coronary' heart disease	total HD	death rate all causes
1921	2229	96	2325	16525(-26% since 1901)
1939	3036(+36%)	1270(+1,222%)	4306(+85%)	16656(+1%)
1951	2172(-28%)	2660(+109%)	4832(+12%)	14314(-13%)
1971	1000(-54%)	4820(+81%)	5820(+20%)	13450(-6%)

Sources: Logan 1951. Registrar General's Annual Reports 1951 and 1971

of the health of middle aged working men to improve in the postwar period.

Whatever gave rise to the postwar discourse on an epidemic of heart disease, which has entered our popular culture with considerable effect, it was not the speeding up of the increase in officially recorded deaths from diseases of the heart. What is obvious from the Table is that the improvement in the health of middle aged men (reflected by the general mortality rate) was getting smaller and smaller. In the 1950s, no-one was going to follow Sir John Simon and take a critical look at industrial conditions. And there was no need, because heart disease had become established as a disease of the affluent working class, a disease of obesity and sloth.

One of the important things to remember when puzzling over the statistics of heart disease is how they are produced. The death rates, by age, by region, by occupation and so on, are calculated using certificates returned by thousands of doctors and hundreds of registrars. They have the task of keeping in order the emotionally fraught events that surround the death of an individual. In the 1920s, when important changes in the rules for filling out and coding death certificates were taking place, there was considerable public concern about two things: hidden murders and 'premature burials'. There was even a Society for the Prevention of Premature Burials. [15] Any death which took place unexpectedly had to be accounted for. And, of course, deaths during working life are always in some sense unexpected ... they take place before a person has reached his or her alloted span, they are more pathological than natural, explanations may be demanded, blame apportioned.

Official health reports in the early postwar period reflect an uncertainty about whether heart disease was (still) on the increase. For example, the Registrar General's Statistical Review for the year 1951:

> The steep rise in diseases of the coronary arteries as a certified cause of death in this and other Western countries has provoked much speculation ... That there has been a change in certification habits cannot be denied. There is evidence too of a tendency at times to mention coronary disease as a terminal development during the course of other illnesses or to use the term as a more 'precise' description of the cause of death of elderly persons...[16]

The Registrar General feels that, notwithstanding all this, it *may* be the case that changes in 'nomenclature' have 'concealed, or even originated from, a real increase in the incidence of the disease, the true magnitude of which must remain uncertain'. The Registrar General's solution to this puzzle and to the inadequacy, as he sees it, of existing vital statistics to answer the question, demonstrates very well how, in certain cases, aetiological theories were used to justify the validity of the very statistics use to 'prove' or to 'test' those theories:

> ... we must try to determine the factors responsible for the occurrence of the disease at the present time — among which diet, mental stress and lack of physical exercise have come under suspicion — and judge whether the varying influence of the causative factor or factors can have produced a rising incidence of the disease.[17]

The logic of this statement is as follows: We think that coronary heart disease has increased because of changes in diet and stress and exercise (because statistics show it to have risen at a time when these other factors may also be changing). But, because we are not too sure of the accuracy of the statistics of the disease, first we need to establish its causes. Then we can see if these causes have increased as well, and judge the accuracy of the disease statistics accordingly. This sort of circular argument attends the entire history of the 'epidemic of heart disease' in various guises.

Another problem noted by the Registrar General's Report for 1951 was that:

> Heart disease described in terms such as coronary, myocardial, degenerative, ischaemic, arteriosclerotic and hypertensive or senile are frequently mentioned on death certificates, but it is evident that the same descriptions, though not used indiscriminately ... are often intended to convey different meanings when used by different certifiers. ... it is becoming increasingly urgent to clarify the nomenclature of these diseases in order that the mortality trends can be properly analysed and those cardiac conditions responsible for premature and preventible deaths distinguished from those, if there are such, that are the inevitable concomitants of old age ... [18]

The Registrar General's Statistical Review for the year 1956 has a whole section devoted to a study of the accuracy of death certification. A total of 1404 deaths had been studied. A clinician had filled out a death certificate before autopsy, and another certificate was filled out by a pathologist afterwards. 'Arteriosclerotic heart disease' was mentioned in 227 of these, but agreement between clinician and pathologist was reached in only 93 cases. It seems that clinicians were tending to overdiagnose heart disease, in relation to what the pathologist 'saw'. The Registrar General cautions that:

> In assessing the cause of the disagreement, great attention had to be paid to the use of words and it appeared that the clinician used words suggesting an acute attack more often than was warranted by the facts as revealed post mortem.

When they came to certificates stating 'myocardial degeneration', the researchers found that this term was used 'vaguely' and much more frequently outside of hospitals than inside them. 'In only one case did the clinician's and the pathologist's diagnoses agree.' The ratio of 'other' myocardial degeneration to 'arteriosclerotic' heart disease in this study was one to ten. In national statistics at that time it was over eight to ten, reflecting what must have been an enormous difference between the attitudes of hospital doctors and general practitioners to the meaning of this term.

The final category of cause of death related to the heart in this study was 'other and unspecified diseases of the heart'. Agreement was reached between clinicians and pathologists in only three out of the thirty-six cases so labelled. Two kinds of disagreements arose, one in which the pathologist preferred a more specific diagnosis such as arteriosclerotic heart disease. The second major source of disagreement was: '... the use of words indicating 'cor pulmonale' resulted in disagreements of assignments between this rubric and those for chronic bronchitis and emphysema .' The Registrar General sums up the findings of this study:

> A vicious circle seems to be operating. The present state of our knowledge of coronary disease is such that accurate diagnosis and classification is not easily achieved, while at the same time the present classification is probably tending seriously to confuse the issues involved. [19]

In 1962, the Chief Medical Officer to the Ministry of Health could still write:

> Cardiovascular disease ... is pre-eminently a disease of old age. Although dependent upon degenerative changes in the heart the final fatal illness is often

> precipitated by inclement weather, epidemics of
> influenza or other infections, so that year to year
> fluctuations are largely a reflection of such epidemics
> in the community, just as the long term trend reflects
> the increasing number of old persons. [20]

In his report for the year 1963, the CMO echoes the concern with the health of middle aged men. Life expectancy at age 55 had deteriorated since 1953, and a spell of cold weather had increased the deaths, 'attributed to coronary artery disease and arteriosclerosis or other myocardial degeneration.' [21]

Somehow the theory of dietary causation has grown up, in ways that this paper has by no means fully documented, on the basis of mortality statistics which are so readily admitted to be influenced by many things other than a concern to pin down the exact cause of death. A 'community study of heart disease' carried out in the 1960s, provides further examples of this problem. The investigators' aim was to trace every case, fatal or non-fatal, in a limited area (Edinburgh) over a period of about a year and thus to establish 'true incidence' of the disease. They found themselves faced with great problems, for example, in deciding who was a real 'case', especially in cases of 'sudden death', because:

> ... by this term, some have meant unexpected death
> either instantaneously or within an hour of the onset
> of symptoms, whereas others meant death within 24
> hours, or even as long as six days.

and that:

> Conventional necropsy alone is a poor method of
> establishing the cause of sudden death ... Existing
> pathological techniques do not reveal myocardial
> infarction with any certainty under about four
> hours. One cannot justifiably conclude from the
> presence of extensive coronary atherosclerosis ... that
> sudden death has been necessarily due to myocardial
> ischaemia. Existing pathological techniques for the
> earliest assessment of myocardial infarction leave a
> lot to be desired.[22]

Another similar 'incidence study' in Tower Hamlets, London, reported some even stranger results:

> ... survivors of possible acute myocardial infarction
> (MI) (i.e. those in whom diagnostic tests were
> inconclusive) re-entered the study with definite

acute MI at the same rate as those who had survived
attacks categorised as definite acute MI.[23]

It should be explained that 're-entry to the study' here means
returning to hospital with symptoms which this time are accepted as a
'real' heart attack (this includes sudden death). In this study, the
factor which discriminated best between those who died of heart
attacks and the healthy population of the area was the presence of pre-
existing chronic illness. One in six of those aged over 55 who died
during the study were not working due to ill health, and one in three of
those aged 55-64. These proportions are six times greater than the
proportion of the disabled in the general population locally. From this
we get a picture of people in middle age, often out of work, usually due
to chronic illness, who die (as the Edinburgh group noted) without
medical assistance despite 'pre-existing symptoms which passed
unnoticed'. [24] Orthodox epidemiology has taken it as given that heart
disease is the sort of disease that kills 'suddenly', although it is admitted
that if all persons with chest pain severe enough to be 'suspicious',
breathlessness and other associated symptoms, which might be
premonitory of a 'heart attack', were admitted to hospital, existing
services would be swamped.

A historical approach leads us to turn this orthodoxy on its head and
ask rather, when and how did it become necessary for 'unexpected'
deaths of persons in the middle and older age groups to be accounted
for by assigning them a specific disease category? A glimpse of the
process is offered by Greenwood's strange little book on death
certification. He complains that:

> A medical certificate of the fact of death may still (in
> 1928) be given, and is constantly given, by a doctor
> who has never set eyes on the presumably dead body
> ... [25]

In 1923 the Society for the Prevention of Premature Burials, to which
Greenwood seems to have been one of the spokesmen, drafted a Bill
which was presented to the House of Commons in June 1924. The Bill
proposed that all death certificates should be given by doctors who had
first been obliged to examine the body. The reaction of the Registrar
General and the Ministry of Health was that this would prove 'too
expensive'. In the ensuing debate in the House, it emerged that 'only
40% of persons buried in Great Britain at this time were certified on
medical evidence actually to be dead'. [26]

Greenwood quoted many coroners' reports from newspapers in
support of his contention that death certificates, far from being an

accurate reflection of the cause of death, were not even an accurate record that death had occurred. For example, this from the Westminster coroner, quoted in the *Sunday Times* of January 14, 1923:

> Doctors . .. commonly certify when they have not
> seen the deceased for months ... these certificates are
> mere guesses at the cause of death, are generally
> given to save trouble and to please the relatives ...
> One case was certified as 'natural death from heart
> disease' when the dead man was found to have a
> dagger through his heart! [27]

These passages are merely some of the most striking amongst a great deal of evidence that unexpected and/or premature deaths became the sort of event that could be neatly dealt with by attributing them to such as an entity as 'coronary' heart disease. Granting death certificates is essentially an administrative process, part of the way in which medicine patrols the borders of social normality and preserves certain images of the social order. And yet, the administrative process generates a set of statistics which are used to try and discover the 'causes' of something which has been thereby constituted as a unitary biological phenomenon.

As people expected to live longer, death became a pathological rather than a normal event at a later and later phase in the life history. As the power of workers' organisations to gain compensation for industrial illness and deaths grew, it became necessary to account for the early death of many workers in heavy and dirty industries. The idea that a miner or a labourer 'wore out' ('degenerated') faster than a vicar or a teacher was no longer just part of the natural order. It needed to be explained in terms of a disease. A contemporary example of this process in action is brilliantly given by Max Atkinson in a book on suicide:

> Towards the end (of the autopsy), the coroner's
> officer asked the pathologist: 'Well, have you found
> anything for me?' ... The pathologist had paused
> after the question and had picked up the heart of the
> deceased which he seemed to be examining closely . ..
> [he] looked up from his examination of the heart and
> said: 'Well, I'd like to give you 'shock' — 'shock' in
> the medical sense that is, because the shock of the
> operation is what really stopped his heart beating,
> but this coroner doesn't like 'shock' does he?' The
> coroner's officer confirmed the this was indeed the

case, to which the pathologist replied: 'I could give
you 'heart failure' then, how would that be?' 'That'll
do me fine,' replied the coroner's officer. [28]

In the past two or three years, the pillars of the medical profession
have trembled slightly under the impact of a growing debate on the
quality of death certification. A Committee on Medical Aspects of
Death Certification was set up jointly by the Royal Colleges of
Pathologists and Physicians, which reported in 1982. Their report
helps us to see how the record-keeping procedures of the practice of
medicine produce 'rates of mortality' which must still be heavily
dependent upon the practical needs of clinical medicine in hospital
and community, rather than upon the needs of epidemiological
research.

> There are ... certain administrative matters which
> allow or even encourage inaccurate certification.
> The certification of death in hospital is usually done
> by the most junior doctors ... furthermore, they may
> avoid any mention of septicaemia or alcohol related
> disorders because they know that if these words
> appear on the death certificate ... the coroner will
> become involved. Many doctors ... are aware that
> certain items on the certificate attract the attention
> of the local registrar and lead to a death being
> referred to the coroner. With this knowledge, they
> avoid the use of such terms. The extent of this
> practice is not known. [29]

'Certain items' in this context include industrial disease.

H.M. Cameron and E. McGoogan, two leading pathologists, one of
whom sat on the joint Committee quoted above, had previously
published a paper on the results of a series of autopsies carried out in a
major British hospital. They concluded:

> There is little doubt that myocardial infarction and
> its abbreviation 'AMI' are commonly used as
> convenient labels when the cause of an unexpected
> death is not known ... in general practice, the
> diagnosis of coronary artery disease is little more
> than guesswork, and it is likely that guesswork is not
> confined to general practice. ... In our experience,
> statistics from death certificates are so inaccurate
> that they are unsuitable for use in research and
> planning. [30]

In Cameron and McGoogan's study, AMI and 'arteriosclerotic heart disease' (ASHD) were two of the most commonly 'overdiagnosed' diseases (i.e. most likely to appear on pre-autopsy death certificates and fail to be confirmed pathologically). They propose that, 'Routine autopsies on cases with these clinical diagnoses' should be carried out to check for the presence of other conditions, particularly rheumatic heart disease (the old disease of poverty), and gall bladder disease and peritonitis (which, as causes of death in hospital, might be described as diseases of neglect).

It is not the contention of this paper that people have been dying of heart disease 'caused' by poverty, industrial hazards or unemployment and other stresses, any more than that their illness has been caused by obesity and sloth. Rather, I mean to question whether we can even go so far as to talk about a modern 'epidemic of heart disease' at all, without a far more careful historical analysis of how death certification and the production of health statistics play a role in maintaining a social order at different periods in time. The pattern of material deprivation, poor working conditions, low pay, persistent chest pain (whether attributed to heart or lungs) and other forms of chronic illness, co-existing with early death, seems to have remained with us. It is illustrated in the social class, occupational and regional mortality differentials, and is commented upon by official government spokesmen on health and independent researchers throughout the period 1900-1980. During this period, ideas about disease entities and their aetiology have undergone major changes.

What can be seen now, in the current health education orthodoxy, as described by Wendy Farrant and Jill Russell [31] is the political importance of a major disease category and of the causal theories which seem to arise 'naturally' and inevitably out of the definition of the disease itself. If we accept the entity 'coronary heart disease', we have already conceded half the case for the existence of an 'epidemic' caused by affluence and declining levels of physical activity. It is not enough to ask 'what are the social factors in the causation of coronary heart disease?' — and thereby leave to the medical profession the major initiative in defining disease categories, whilst accepting the ancillary task of slipping in a few social or historical bits here and there. The historical evidence demands that we suspend the assumptions behind conventional diagnostic categories and locate, not just the 'causes', but also the diseases themselves in their social and historical context.

This paper could not have been written without many discussions with colleagues and friends which took place before and during the process, particularly Karl Figlio, David Blane and Norman Weinstock. I am also very grateful for detailed comments and extensive encouragement and guidance from Celia Davies.

Notes

1   J.A. Ryle and W.T. Russell, 'The Natural History of Coronary Heart Disease — a Clinical and Epidemiological Study', *British Heart Journal* , vol. 11(1949), pp. 370-389.

2   M. Oliver, 'Does Control of Risk Factors Prevent Coronary Heart Disease?', *BMJ*, vol. 285 (1982), pp. 1065-1066.

3   Ministry of Health, *Annual Report of the Chief Medical Officer 'On the State of the Public Health' for the Year 1922* (HMSO, 1923), p. 15.

4   Ibid., p. 16.

5   Ibid., p. 16.

6   *The Registrar-General's Statistical Review of England and Wales for the Year 1925* (HMSO, 1927), p. 55.

7   *The Registrar-General's Statistical Review of England and Wales for the Year 1926* (HMSO, 1928), pp. 85-86.

8   *The Registrar-General's Statistical Review of England and Wales for the Year 1930* (HMSO, 1932), pp. 67-68.

9   Ministry of Health, *Annual Report of the Chief Medical Officer for the Year 1932* (HMSO, 1933), p. 17.

10   Ministry of Health, *Annual Report of the Chief Medical Officer for the Year 1933* (HMSO, 1934), p. 206.

11   W.P.D. Logan, 'Mortality in England and Wales', *Population Studies*, vol. 4 (1950), pp. 132-178.

12   Ministry of Health , *Annual Report of the Chief Medical Officer for the Year 1933* (HMSO, 1934), p. 252.

13   Charles Webster, 'Healthy or Hungry Thirties?', *History Workshop Journal*, no.13 (1982), pp. 110-129.

14   Sources: Logan, 'Mortality', and *Registrar-General's Statistical Reviews* for the years 1951, 1961, 1971.

15   Sir George Greenwood, *The Law of Death Certification* (Published on behalf of the Society for the Prevention of Premature Burials, 1928).

16   *The Registrar-General's Statistical Review for the Year 1951* (HMSO, 1954), p. 218.

17   Ibid., p. 218.

18   Ibid., p. 218.

19   *The Registrar-General's Statistical Review for the Year 1956* (HMSO, 1958), pp. 188-191.

20   Ministry of Health, *Annual Report of the Chief Medical Officer for the Year 1962* (HMSO, 1963), p. 25.

21   Ministry of Health, *Annual Report of the Chief Medical Officer for the Year 1963* (HMSO, 1964), p. 23.

22   M. Fulton et. al., 'Sudden Death and Myocardial Infarction', *Circulation*, Suppl.4, vols. 39-40 (1969), pp .(iv)182-(iv)193.

23   H. Tunstall-Pedoe et. al., 'Coronary Heart Attacks in East London', *Lancet* (1975) (ii), pp. 833-838.

24   Fulton et. al., 'Sudden Death', pp. 189.

25   Greenwood, *Death Certification*, p. 14.

26   Ibid., p. 14.

27   Ibid., p. 20.

28   J.M. Atkinson, *Discovering Suicide: Studies in the Social Organisation of Sudden Death* (London, 1978), p. 98.

29   Joint Committee of the Royal Colleges of Physicians and Pathologists, 'Medical Aspects of Death Certification', *Journal of the Royal College of Physicians of London*, vol. 16 (1982), pp. 205-218.

30   H.M. Cameron and E. McGoogan, 'A Prospective Study of 1152 Hospital Autopsies. I: Inaccuracies in Death Certification', *Journal of Pathology*, vol. 133 (1982), pp. 273-283.

31   Wendy Farrant and Jill Russell, *'Beating Heart Disease': A Case Study on the Production of Health Education Council Publications* (Institute of Education Health Education Publications Project, unpublished monograph, 1983).

Part Three: Compensation

9 THE RISE AND DECLINE OF WORKMEN'S COMPENSATION [1]

Peter Bartrip

Victorian Britain can be seen in terms of a largely full-employment but low-welfare society. In such circumstances a workman's greatest need was for a fit and healthy body, for only with such could he expect to perform the work required to obtain for himself and his family food, shelter and clothing without recourse to the poor law, private charity or other forms of non-wage financial support. One of the major threats to bodily and material sufficiency, especially for miners, railwaymen, merchant seamen and others in dangerous employment, was provided by industrial injury or disease without compensation. [2] It was not until 1897 that Parliament passed a Workmen's Compensation Act giving large groups of workers, in the event of physical injury, a statutory right to compensation from their employers regardless of the employer's fault and largely regardless of their own part in precipitating their misfortune. This Act established what Beveridge was later to call 'the pioneer system of social security'. [3] Its operation will be one focus of this paper, but first we may look at the position of victims of occupational injury before the passing of that Act.

I Before Workmen's Compensation

Until 1837 there was no reported High Court case of an employee suing his master for damages as a result of injury negligently sustained at work. But the precise significance of this fact is not altogether clear, for suits may have been decided by local courts. This, indeed, is quite likely to have been the case. By tracing back the first High Court case, that of *Priestley v. Fowler*, [4] to its place of origin, rural Lincolnshire, we find nothing in local newspaper reports to suggest it was unique. [5] Furthermore, we know from other legal evidence that by the 1830s the idea that a master owed a duty of care to particular individuals in his employ, for example, apprentices, was well established. [6] All this being so, we are faced with the question of why it was as late as 1837, by which time the process of industrialisation was well advanced, before a personal injury action by an employee against his employer reached the High Court. It surely was not the case that work injury was new, for common sense tells us that industrial and agricultural accidents are as old as mankind itself. Part of the answer may be provided by an understanding of the paternalistic nature of pre-industrial society, in which production was quintessentially an activity of small units in which a master shared labour with workers who were often members of his own household. In such circumstances it is easy to see that the prevention, treatment and compensation of accidents and accident victims occurred on a personal level without recourse to the courts. It is a truism that under the pressures of industrialisation this social relationship broke down to be replaced with one dominated by class conflict and the cash nexus.[7] In these circumstances industrial injury became a matter for the Courts, for Acts of Parliament and government inspection. Yet even this interpretation fails to explain why *Priestley v. Fowler* occurred so 'late', for most historians would agree that by the 1830s industrialisation had long revolutionised English society. Thus, in order further to account for the absence of case law we must point either to the problems encountered by workmen in suing their employers or to the continuing charity of those employers when faced with labourers injured in their service.

There can be no doubt that to commence legal action against an employer was a very serious step for a workman to take. Unless supported by a wealthy patron (in the 1840s Lord Ashley financed several employers' liability suits), an action was likely to be ruinously expensive; it could also mean dismissal with the certainty that no other master in the district would ever hire the 'troublemaker' again.[8] Furthermore, given the social composition of those hearing cases — middle to upper-class judges and a jury selected on a property qualification — the chances of success were limited.[9] At the same time, it is unrealistic to suppose that all employers were totally hard-hearted in their attitudes towards injured employees. Private philanthropy and charity did not die with the establishment of industrial capitalism; Charles Dickens's Bounderby and Scrooge were not typical examples of nineteenth-century businessmen. Hence a realistic appraisal of why there were no cases before *Priestley v. Fowler* must emphasise both the difficulties facing the injured and the likelihood that many masters made at least some reparation to injury parties.

Priestley v. Fowler was not followed by an avalanche of litigation. Indeed, it was not until the 1850s that there occurred an action arising from the new conditions of industrial life (the Priestley case, which was lost by the plaintiff on appeal, arose from injuries sustained through the over-turning of a butcher's cart). From the 1850s, as more cases arose, the common law underwent considerable modification. The effect of much of this was detrimental to the chances of plaintiffs securing judgment. In particular, in the decades following *Priestley v. Fowler,* the courts developed three so-called legal 'fictions' : common employment, contributory negligence, and *volenti non fit injuria* which, together with the continuing problem of financing legal actions, combined to render employers' liability for personal injury virtually a dead letter. We may now consider briefly the significance of the three 'fictions'.

Lord Abinger, when giving judgment in the Court of Exchequer on *Priestley v. Fowler* stated, not altogether accurately, that there was 'no precedent for the present action by a servant against a master'. He went on to say that his decision would, therefore, be based upon 'general principles'. The general principle influencing his finding for the appellant, Fowler, the butcher, was that to decide the other way would be to 'open the floodgates' of litigation with 'alarming' consequences. Such policy, rather than legal considerations, informed much judicial thinking over the early and mid-Victorian period. Thus, building on the judgment in *Priestley v. Fowler* and a later American case, *Farwell v. Boston and Worcester Railroad Corporation* ,[10] the Court of Exchequer introduced the doctrine of common employment to English soil in *Hutchinson v. York, Newcastle and Berwick Railway Co.*(1850).[11] The judgment held that no action could lie if a work injury or fatality were caused by a person in common employment with the injured party, that is, a fellow-workman. As the scope of common employment was later extended, the rule came to mean that even managers were deemed to be in common employment with manual labourers. Before this rule was abolished by the Law Reform (Personal Injuries) Act 1948, it excluded many injured workers from damages.[12] The same was true of the defence of contributory negligence, whereby no action could succeed if it were shown that the accident victim was to any degree responsible for his fate. This defence was not amended until 1945.[13] The third 'fiction' was that of *volenti non fit injuria* which was the principle that in accepting dangerous employment a workman willingly consented to the risks and implicitly renounced all claim to be compensated. Hence, for an action to succeed it became necessary for the plaintiff to demonstrate that his injuries were wholly and directly the fault of his master and incurred in circumstances excluding the defence of *volenti*. In an economy increasingly dependent on a complex division of labour in large units of production this was very difficult to achieve. Accordingly, only a few successful actions occurred before amendment of the law in 1880. Of the above barriers only contributory negligence applied outside the workplace. Thus, while successful actions by injured railway passengers against railway companies were relatively commonplace, it was virtually unknown for an injured railway servant to secure damages.[14]

An abortive attempt to abolish common employment was made in 1862, but sustained reforming activity only got off the ground with the passing of the Second Reform Act (1867), the foundation of the Trades Union Congress (TUC) (1868) and the election of trade unionists to Parliament (Alexander Macdonald and Thomas Burt in 1874). During the seventies a number of Bills, mostly aimed at limiting or abolishing the common law defences against employers' liability actions, came before the House. A Select Committee investigated the question which became an election issue in 1880.[15] Throughout this long period of public debate employers and their representatives claimed that a change in the law would mean disaster for British industry. Their adversaries, on the other hand, claimed that they only wanted working people to be put on an equal footing with the rest of the community. In 1880 the hopes and fears of the conflicting parties seemed to have been realised when Gladstone's newly elected government secured the passage of the Employers' Liability Act.[16]

For all the intense debate over much of the preceding decade and for all the extravagant claims, pro and anti, made for the various Bills of the 1870s, the 1880 Act represented only a minor adjustment to, rather than a revolution in, liability law. It abolished the doctrine of common employment in only four specific circumstances; the Act did not apply to seamen, domestic servants or, with the exception of railway servants, non-manual workers; a ceiling of three years' wages was placed on damages recoverable under the Act, a particular hardship in the case of minors; the defences of *volenti* and contributory negligence remained. Furthermore, in some localities the significance of the Act was further restricted by the widespread practice of contracting out, whereby workers signed away their right to take advantage of the Act.[17] Since the necessity to prove fault remained and since nothing was done to help finance actions or redress the inequality of plaintiffs and defendants, it is clear that the 1880 Act was no more than a moderate reform. Consequently, almost as soon as it became law, the movement to amend it began.

During the eighties and nineties a rash of Employers' Liability Bills, many drawn up by the TUC, came before Parliament. Most of these Bills tinkered with the principal Act by proposing such reforms as the abolition of contracting out, further erosion of common employment, the raising of the maximum level of compensation, or the inclusion of further categories of employee. They failed because a Parliament in which the working man continued to have a limited voice was unwilling to re-open a question considered to have been settled in 1880. The objective behind these reform attempts was to place a bigger and more certain financial burden upon more employers in order, partly, to compensate the injured, but also to increase financial incentives to adopt safety precautions. However, given that most employers insured their risks, there was little likelihood that the reformers' safety intentions could be realised. Although industry was becoming safer throughout the nineteenth century, there is little evidence to suggest that amendment of liability rules played much part in the process.

In 1893 Herbert Asquith, Gladstone's Home Secretary, introduced an Employers' Liability Bill in the Commons. When it came up for its second reading Joseph Chamberlain moved an amendment which was to revolutionise developments in industrial compensation:

> . . . no amendment of the law relating to Employers'
> Liability will be final or satisfactory which does not
> provide compensation to workmen for all injuries
> sustained in the ordinary course of their employment
> and not caused by their own acts or default.[18]

Chamberlain's point was that even a reformed employers' liability law would leave uncompensated all those whose injuries were caused by 'act of God'. If employers were to be held liable for accidents over which they had no personal control, it was fair and logical that all victims who were not themselves negligent should gain redress. Such a proposal, calling for workmen's compensation on an insurance rather than a negligence basis, was not entirely original; it harked back to insurance ideas raised in the debate over the 1880 Act and proposals for reform made by Edwin Chadwick in the 1830s and 1840s.[19] The difference was that Chamberlain's amendment yielded a legislative consequence, the Workmen's Compensation Act, 1897.

Before this Act was passed there was, of course, a great deal of parliamentary and extra-parliamentary debate, much of it involving the conversion of hostile parties. Even in 1897 the TUC and individual unions remained critical of the insurance solution.[20] Why this was so is arguable. The unions themselves claimed to oppose any measure which removed the costs of accidents from the employer, thereby also removing safety incentives. Their critics, on the other hand, argued that it was unrealistic to suppose that employers were bearing the costs of a significant proportion of accidents. They maintained instead that unions favoured tort law because it safeguarded their own interests both by nurturing industrial tension and by constituting a powerful recruiting device (unions made great play of the legal and financial assistance they rendered to injured members). If the unions were unenthusiastic about workmen's compensation, so too were employers for they were gaining a potentially huge overall liability to pay compensation regardless of fault. Ultimately they were readier to accept this than the alternative, that is, a revamped employer's liability law whereby they might incur the obligation of paying large and unpredictable amounts of damages. The merit of workmen's compensation, in their eyes, was that it would involve relatively small individual sums for which budgetary provision could be made. But it is significant that workmen's compensation was backed by no powerful interest groups — it was a compromise measure. This point goes some way towards explaining the broad based hostility aroused by the Act in operation.

II The Workmen's Compensation Act

The Act was extremely complex, but essentially it stated that: 'If in any employment to which this Act applies personal injury by accident arising out of and in the course of employment is caused to a workman, his employer shall . . . be liable to pay compensation . . . ' regardless of his fault in causing the accident, and, largely, regardless of the part of the workman in bringing about his injury. The places of employment to which it was applicable were defined as 'on or in or about a railway, factory, mine, quarry, or engineering work, and . . . on, or in or about any building which exceeds thirty feet in height and is either being constructed or repaired by means of a scaffolding, or being demolished, or on which machinery driven by steam, water, or other mechanical power, is being used for the purpose of the construction, repair, or demolition thereof'. Compensation, which was payable from two weeks after the accident, was to be based on previous earnings with a ceiling of £300 for a lump sum in the event of death, and up to £1 per week in cases of total or partial incapacity.[21] The Act came into effect on 1 July 1898.

The new Act was greeted in emotional terms by its supporters and opponents. Some employers felt it would cripple the international competitiveness of British industry despite the fact that other countries, notably Germany, already had similar legislation. On the other hand, its champions considered it to be 'a great boon' to workers. One historian terms the Act 'revolutionary' and has even gone so far as to claim: 'There can be no doubt that the Workmen's Compensation Act was one of the most important ever passed by Parliament'.[22] This is certainly claiming too much for a measure which (until 1906) applied to only a minority of the country's total workforce, imposed relatively low ceilings on maximum awards and relatively complex regulations on eligibility for benefit. However, the Act was of significance in one practical and one symbolic way. It established work injury victims as an elite group, analogous to war pensioners, eligible for special benefits denied to other, perhaps similarly injured, members of society. This position was consolidated as later legislation substantially enlarged the size of the elite. Moreover, even after the substitution, in 1946, of the Industrial Injuries Scheme for the workmen's compensation system, the preferential treatment of work injury victims was retained. The symbolic importance of the 1897 Act was in conferring a working-class right — comparable with the right to vote or to form trade unions — which, if it did not remove the longstanding disadvantages of the negligently injured worker relative to those injured outside the workplace, promised to circumvent them. Hence, symbolically, the Act constituted another step towards workers' achievement of social, legal and political parity with fellow citizens. Whether the Act's practical impact warranted the hopes and fears of those who witnessed its passing is another matter. The next two sections consider the scope and significance of the workmen's compensation system in operation.

III The System at Work: Dimensions

It is not easy to determine the scale of workmen's compensation during its early years owing to the severe shortcomings of the statistics. For this reason and to avoid the interruption of the war years, during which no statistics were collated, attention here will be confined to the figures for the inter-war years, principally for the seven industries (factories, mines, shipping, docks, quarries, railways and construction work) which were required by a 1907 Home Office order to compile detailed workmen's compensation statistics. Since these seven industries accounted for a large proportion of compensation cases and payments, much may be surmised from their statistics about the overall dimensions, costs and effectiveness of workmen's compensation.[23]

By the end of the 1930s (see Table 1) over 450,000 compensation cases were each year generating payments of close to £7 million for accident victims and their dependants. This meant that every year some 5 to 6.5 per cent of the workforce in the relevant industries were making a *successful* claim for compensation. Hence, on average there was a strong possibility that in the course of a working life

Table 1 Statistics of Compensation in the Seven Industries, 1919-1938

Year	Fatal	Non-Fatal	Total
1919	3,293	365,176	368,469
1920	3,531	381,986	385,517
1921	2,385	283,361	285,746
1922	2,489	390,423	392,912
1923	2,657	477,378	480,035
1924	2,878	487,442	490,320
1925	3,030	473,055	476,085
1926	2,345	368,563	370,908
1927	2,567	455,852	458,419
1928	2,735	461,485	464,220
1929	2,819	478,602	481,421
1930	2,621	458,509	461,130
1931	2,315	396,571	398,886
1932	2,011	364,864	366,875
1933	2,072	359,971	362,043
1934	2,229	401,459	403,688
1935	2,640	422,699	425,339
1936	2,286	459,271	461,557
1937	2,370	486,495	488,865
1938	2,498	456,725	459,223

every eligible workman, and especially those engaged in the dangerous trades, would have recourse to workmen's compensation at some time. In fact, aggregates of benefits and cases tend to minimise the dimensions of the system in two ways. Firstly, case numbers refer to numbers of accident victims; the annual number of beneficiaries in any single year clearly exceeded one million if dependants are taken into account. Secondly, published figures for payments refer only to direct benefits; they exclude legal, medical and administrative costs, insurance company profits and money paid by insurers into their reserves. There is no way of ascertaining precisely how much these additional factors would have inflated overall costs. The Home Office did, however, publish rough estimates of the total cost of compensation in the seven industries (Table 2) and, from 1930, of all industries (Table 3) , but these provide no more than a general indication of overall expense and should not be relied upon for indications of trends.

Table 2 Estimates of Total Costs in the Seven Industries, 1919-1938 (£m.)

Year	Cost	Year	Cost	Year	Cost
1919	6.0	1926	7.85	1933	7.0
1920	8.5	1927	8.25	1934	7.0
1921	8.5	1928	8.25	1935	7.5
1922	8.5	1929	8. 5	1936	8.0
1923	9.0	1930	8. 0	1937	8.0
1924	9.0	1931	7. 5	1938	8.5
1925	9.0	1932	7. 0		

Table 3 Estimates of the Overall Costs of Workmen's Compensation, 1930-1938.

Year	Estimate (£m.)
1930	12.0
1931	11.0
1932	10.5
1933	10.0
1934	11.0
1935	12.0
1936	12.5
1937	13.0
1938	13.5

The fundamental question arising from the statistics presented in these Tables is whether they should be considered large or small. It is impossible to provide an objective answer, for any conclusion must rest upon assumptions of magnitude relative to other indices of expenditure or some notion of what burden society could or should afford. Some observers looked at the plight of impoverished families stripped of their breadwinners and concluded that compensation should be greater; others looked at struggling industries and claimed that a contributory cause of British industrial decline was the extent of expenditure on social services relative to that of overseas competitors . Naturally, the division of these views tended to be between employers and workers, trade unions and employers' organisations, Labour and Conservative Parties. The Balfour Committee on Industry and Trade, appointed by Ramsay MacDonald in 1924, looked into the question of industrial costs as part of its inquiry into factors affecting industrial and commercial efficiency. Its final report concluded that although social service expenditure was under one per cent of the gross value of output and therefore relatively insignificant it 'may in certain cases be a very appreciable element in costs, and there is every reason for continued vigilance to ensure the maximum of economy'.[24] Compared with other social service expenditure, the overall scale of workmen's compensation costs was not great. But workmen's compensation was the only scheme financed entirely by employers. If there were grounds for criticising its adequacy in individual cases, as there were, this was little consolation to the scheme's paymasters in times of national economic difficulty.

IV The System at Work: Problems and Issues

There can be little doubt that the 1897 Act failed to measure up to the extravagant claims made for it at the time of its passing. One of the principal shortcomings was the level of benefits payable. Average weekly payments of some 11s per week in 1902 (average wages were then about 26s 6d) were modest indeed. In relation to living costs the value of benefits, which was subject to constant criticism, especially by the trade unions and the Labour Party, improved little over the years. There are innumerable examples of such criticism, even from such unimpeachably objective sources as the *Law Times* which, in 1928, called benefit scales (revised by Parliament in 1923) 'tragically inadequate'.[25] Looking back on this issue in 1945, a solicitor, in evidence to an official committee, claimed that 'the fundamental vice of the Workmen's Compensation Acts was the misery and injustice inflicted upon the injured workmen by the grossly inadequate weekly payments'.[26] A working man's view of benefits is provided by the evidence of a Canadian carpenter submitted to the Nuffield College Social Reconstruction Survey in the early 1940s: 'I have searched my dictionary from the front to the back to find a suitable word to express the opinion of the consumer and failed to find it'.[27]

When it is remembered that workmen's compensation benefits were based primarily upon previous earnings and took little or no account of needs, the significance of the actual sums paid to beneficiaries becomes clearer.[28] A father of six children who was injured at work could expect to receive no more in compensation than a similarly incapacitated married, yet childless, man. The financial consequences of industrial disability could, therefore, vary greatly. Family size was not, however, the only variable affecting the financial needs of the injured; the nature of the injury was a further factor. Yet workmen's compensation benefits took no account of hospitalisation, other medical treatment or the necessity for equipment such as artificial limbs. Clearly, recourse to any of these facilities could prove a severe drain on a worker's resources. Although a trade union might provide legal expertise to the minority of workers who were union members, an injured employee's expenses arising from his injury could pose severe financial difficulties before the problem of income replacement arose. Of course, workmen's compensation was not intended to provide in full for the financial loss sustained by work accident victims; it was not even meant to compensate in full for lost earnings (and the Act provided nothing for pain and suffering, and disfigurement). It was intended to share the loss occasioned by an accident between employer and employee; but given that workers sustained the physical loss as well as a financial loss, it is arguable that the share-out was inequitable.

Inadequate compensation was only one grievance of those who succeeded with a claim under the Acts. Another was the delay in obtaining a settlement. This could take up to one year; for a family stripped of its breadwinner, possibly with few savings and, perhaps, also incurring substantial expenses arising from the injury, such delays could be a very serious matter.[29] More serious still was the absence of any security that payments would be maintained. The 1897 Act permitted claims by those injured under circumstances specified in the statute and laid down scales of payment and procedures for securing settlement. It did not, however, guarantee benefits. If an employer or an insurer went out of business, those in receipt of weekly compensation payments became creditors and ran the risk of losing most or all of their regular entitlement. Parliament and various governments made efforts to improve the financial security of beneficiaries by giving them a prior claim on assets over all other creditors, by regulating the insurers and by making insurance compulsory in the coal industry.[30] The effect of these changes was, however, marginal and uncertainty about benefits continued to be a criticism of the workmen's compensation system down to the 1940s. The numbers who lost benefits may not have been great in relation to the total number of beneficiaries, but their plight attracted considerable attention since it highlighted a major defect in the legislation, namely that a legal right conferred upon a particular group could on occasion be literally worthless.

Shortcomings in scales of benefit, rapidity and certainty of payment were, in a sense, secondary problems of workmen's compensation, for those who experienced them had at least succeeded in pressing a claim. Before reaching this stage the claimant and his advisers had to negotiate their way through a maze of ambiguous legislation, contradictory judicial decisions and the risk of litigation. In the early years of the twentieth century few legal authorities referred to the Workmen's Compensation Act in complimentary terms. For Parsons and Bertram it had 'singularly drawn provisions'.[31] According to Lord Brampton

> The whole statute is full of incongruities. In it so many things are said which could not have been meant, and so many things which must have been meant are left unsaid, that one often has great hesitation in even framing a conjecture as to what may have been the views and intentions of its framers.[32]

Many problems derived from the attempt, largely abandoned in 1906, to restrict the scheme to certain groups of workers while excluding others. Hence, the 1897 Act specified that compensation could be claimed by accident victims in the building trade when their accident occurred on 'any building which exceeds 30 feet in height, and is either being constructed or repaired by means of scaffolding', but not by other building workers. This led to a host of disputes regarding the nature of scaffolding and the meaning of the 30 feet rule (for example, could a well which was more than 30 feet deep be considered a building in excess of 30 feet high? What was the position with buildings intended to exceed 30 feet, but of lesser dimensions at the time of the accident? In measuring a building should foundations and/or chimneys be included? How was a building on sloping land to be treated? Could a chair or crawling board be considered scaffolding within the meaning of the Act?).[33] While the 1906 Act removed many anomalies and arbitrary distinctions of this kind, it left others, including the notorious phrase: 'arising out of and in the course of employment'. As Lord Loreburn said in *Kitchenham v. Owners of S. S. Johannesburg*, the words of this clause admitted 'of inexhaustible varieties of application according to the nature of the employment and the character of the facts proved'.[34] This and other clauses were still providing problems of interpretation and much lucrative work for lawyers in the 1930s. But how much litigation did occur under the Workmen's Compensation Acts?

One of the objectives of the framers of the 1897 Act had been to remove compensation from the purview of the courts, the aim being to enable workmen to recover without the expense and difficulty of legal proceedings necessitated by the obligation to prove fault. The expectation was that settlements would be agreed informally between employers and employees with arbitration (which could take place in the County Courts) occurring in only a minority of instances. The settlement of claims was much more contentious than had been somewhat naively anticipated. This was partly because of the policies of the insurance companies (see below), and partly a result of poorly drafted legislation and inconsistent County Court decisions; but more basically it was because a workmen's compensation claim involved complex negotiations over money between the two sides of industry in the emotional circumstances of death or injury. Just how contentious settlement of claims was, however, was itself a matter of contention. Initially this was because the statistics failed to provide an accurate guide to the problem. In these circumstances the Act gained a reputation, based upon accusation, for encouraging litigation. The literature, ranging from the popular press to specialist periodicals, law reports, law texts and *Hansard* bristles with such references as 'flood of litigation' ,'nothing but litigation' and 'more litigation than any other Act which has been passed in recent times'.[35] In 1935 Professor William Robson could still refer to the 'immense volume of litigation' :

> The truth of it is indisputable. In the space of less
> than forty years the legislation which was specially
> designed to give the workman a simple, cheap and

> easy remedy for his injuries has become encumbered
> with an ornate, seductive, writhing mass of case law
> which suffocates the whole scheme by loading it with
> expense delay and difficulty. [36]

Historians and lawyers who have waded through Butterworths'
Workmen's Compensation Cases are likely to sympathise with Robson's
viewpoint.[37] However, the Home Office line was to minimise the
extent of litigation by emphasising that relative to the number of
mutually agreed settlements it was small.[38] This was reasonable so far
as it went — though even a relatively small amount of litigation might
be viewed as excessive in a no-fault system — but it ignored the
possibility that many contentious cases were settled by informal
agreement on terms unfavourable to workmen owing to imprecisions
in the legislation and the prohibitive costs, especially for non-unionised
labour, of going to court. The litigation problem persisted throughout
the existence of the workmen's compensation system.[39] It soured
industrial relations, pushed up costs and ultimately provided one of the
most powerful arguments for changing the system.

If litigation and scales of benefit were two of the major criticisms of
workmen's compensation, the third was the insurance system. When
the 1897 Act became law it was assumed that most employers would
seek insurance to cover their liabilities. This would protect them from
the financial burdens of large claims as well as safeguarding the rights
of claimants who might be deprived of benefit if their employer was
unable to pay compensation. The Act introduced no compulsion to
insure; it was simply assumed that employers would do so in their own
best interests. But claimants had no protection if a small uninsured
employer found himself unable to meet compensation payments, or if
his insured employer, large or small, became incapable of maintaining
insurance premiums. Consequently, the Act could be said to have
fallen between two stalls. It failed, as we have seen, to guarantee that
benefits would be paid in respect of legitimate claims. At the same
time, by allowing insurance there was the possibility that any influence
which the Act might exert in the direction of improving safety via
economic deterrence would be undermined as the employer's liability
became to some extent the insurer's liability.

There are few data on the extent to which claims were lost through
bankruptcies. Some were gathered during the early 1930s in response
to concern about what was happening in the coal industry. Though
their accuracy was disputed at the time, they suggested that between 1
January 1927 and 30 November 1933 280 colliery insolvencies

occurred. In these cases an unknown number of workmen experienced temporary deprivation of benefit. In 24 cases permanent losses were sustained, while in a further 11 the possibility of such losses remained. Thus, at least 245 insolvencies involved no permanent loss. But it was estimated that the 24 confirmed cases involved losses for some 1,500-1,700 claimants exceedings £170,000.[40] These were not insignificant figures; moreover even temporarily lost compensation could involve considerable hardship. As a result of public concern over the sufferings of uncompensated colliers an Act of 1934 made workmen's compensation insurance compulsory in the coal industry. Such insurance would remain valid, notwithstanding failure to pay premiums, provided that a claim arose when the colliery was insured.[41]

While the 1934 Act was significant in terms of principles, its effects were confined to one industry alone. The question of non-payment of compensation by uninsured insolvent employers continued to be raised in Parliament intermittently for the remainder of the thirties.[42] Furthermore, compulsory insurance, demanded in some quarters virtually since the 1897 Act had become law, remained one of the main workmen's compensation reform priorities of the trade unions and Labour Party into the 1940s. Of course, if compensation insurance were to be obligatory there was a strong case that it should be economic, fair and efficient and many voices claimed that the system which emerged after 1897 was none of these. In 1912, at which time the insurance companies were responsible for some 33 per cent of workmen's compensation payments, only 63 per cent of premium income was going towards benefits, the rest was financing commissions, management expenses and profit. Moreover, of the 63 per cent an undisclosed amount consisted of legal and medical expenses. The Holman Gregory Committee found, in the early twenties, that only some 48 per cent of premiums were being paid in benefit while profits exceeded 20 per cent. This prompted the recommendations that there should be government supervision of premium rates, and that at least 70 per cent of premium income should be paid out in benefits.[43]

A Home Office minute of 1922 recognised that companies' premium rates were a 'scandal'.[44] In the following year the government reached agreement with the Accident Offices Association, to which a large number of companies undertaking workmen's compensation business belonged, to the effect that compensation payments would be not less than 5 per cent of premium income, rising to 62.5 per cent in 1926.[45]

The Times feared that the agreement threatened insurance companies' profitability.[46] However an arrangement which allowed 37.5 per cent of premium income to finance operating costs (exclusive of medical and legal expenses, which continued to be categorised as compensation) and yield a profit cannot be accounted particularly severe. In the mid-thirties Professor William Robson could continue to attack the insurance companies for frittering away premiums. Indeed, he blamed them for much of the 'human suffering, economic waste and social injustice' which he associated with the workmen's compensation system.[47]

At first sight it would appear that there was much justification for charges such as Robson's against the insurance companies. Furthermore there is some evidence that the companies engaged in unethical and unscrupulous practices, including inducing recently injured workers to sign away all future claim to compensation in return for relatively small cash payments. But it has to be recalled that insurance companies controlled less than one-quarter of the market (22 per cent of payments and shrinking in 1932). Most employers were in mutual associations (44 per cent of payments and growing) or carried their own risks (34 per cent of payments). In several industries, most notably railways, shipping and mining, the companies' share of the market was negligible. It is, therefore, unrealistic to blame the insurance companies for all the ills of the workmen's compensation system. Mutual association was the more typical form of insurance, and here, overheads were very much lower — that is, some 20 per cent or less of premium income — than for the proprietory companies. Yet the latter were widely assumed to dominate the market.

In the early stages of his Social Insurance and Allied Services Inquiry Sir William Beveridge appears to have shared the notion that all workmen's compensation insurance was wildly expensive and wasteful. Further investigation led him to doubt whether this was the case. In 1942 he found that the proprietory companies were responsible only for some 15 per cent of all compensation payments, whereas mutuals covered some 70 per cent of the market. He also came to see that there were explanations other than profiteering or inefficiency for the companies' high expenses, for whereas mutuals covered big employers in common industries, the commercial companies dealt with many small employers in a wide variety of trades (including employers of domestic servants). In July 1942 Beveridge wrote: 'The expense of insurance with commercial companies cannot well be used as a principal argument against the present system of workmen's compensation. There are many stronger arguments'.[48] Nevertheless, misconceptions about workmen's compensation insurance were so widespread that it is probable that they exerted considerable influence in bringing about reorganisation of the workmen's compensation system.

Safety was a prime consideration of the framers of the 1897 Act, who believed that by imposing the costs of accidents upon employers a more safety conscious attitude would emerge. The danger was, however, that insurance would exert an adverse effect upon industrial safety by imposing the direct cost of compensation upon the insurer rather than the employer. Such a consequence might have been avoided by means of merit rating, that is, the variation of premiums according to accident prevention steps taken by the employer. Such a practice was commonplace overseas. However, the Holman Gregory Committee found that in Britain insurers made little use of a rating system likely to encourage employers to introduce positive safety measures.[49] The conclusions of this Committee were similar to that of an earlier official inquiry which had reported that: 'No evidence has been brought before us which enables us to find any improvement in the direction of safety is to be placed to the credit of this Act'.[50] Although Beveridge considered that workmen's compensation insurance facilitated accident prevention by allowing premiums to be adjusted to ascertained risk, there was virtually no evidence to support this argument.[51] Indeed, the government subsequently discounted Beveridge's view.[52] In safety terms the Workmen's Compensation Acts must be judged to have failed.

V Conclusion

Beveridge listed nine disadvantages in the workmen's compensation system; there were certainly more than can be discussed in this chapter, but inadequate and uncertain benefits, heavy litigation, unsatisfactory insurance, and lack of safety incentives were (rightly or wrongly) the main charges against the system. These channelled into the overriding deficiency of workmen's compensation, namely, its adverse effect upon industrial relations. The system was a conflict-oriented one in that it pitted employer and employee against each other over money matters. This conflict was institutionalised by the involvement of trade unions, employers' associations and insurers with the effect that compensation negotiations sometimes resembled battlegrounds between labour and capital, with each side represented by its battalions of professionals — the doctors and lawyers. One should not over dramatise this; it is important to remember that thousands did receive benefit with relatively little difficulty or delay. If this was inadequate in terms of meeting victims' needs or returning them to their pre-accident financial status, it should be emphasised that the workmen's compensation system was a great advance on the pre-1897 position. Of course, as expectations rose so dissatisfaction with this sort of rationalisation increased. But the crucial point is not that workmen's compensation was a flawed system — few would have disputed this — but that it was widely believed to have been much worse than was actually the case. Criticism of the wasteful insurance system provides a good illustration of this point for, as we have seen, the profligate companies actually held only a small portion of the insurance market. Nevertheless, belief, even if mistaken, may be a powerful force for political change.

It took the Second World War, however, to convert long-standing dissatisfaction with workmen's compensation into a coherent programme of reform. As a result of the national priorities generated by warfare, namely social solidarity and the promise of a better future following the national sacrifice, the replacement of the workmen's compensation scheme by an alternative based more upon co-operation than conflict ultimately became imperative. In these circumstances workmen's compensation was scheduled for abolition — first by Beveridge — and substitution by a contributory social insurance scheme. It is notable, however, that in this process, which cannot be discussed here, the notion of a preferential system for workers persisted, notwithstanding Beveridge's initial desire to establish a uniform system, and also notwithstanding the argument — largely accepted in the social insurance reforms of the 1940s — that benefit schemes should have more regard to the needs of recipients rather than the causes of their misfortunes. The main explanation for this is that the trades unions were unwilling to see members lose a right secured some 50 years earlier, while neither Beveridge nor the Coalition or post-war Labour governments were prepared to confront the unions on this issue.

Notes

1   Much of the early part of this paper is based upon P.W.J. Bartrip and S.B. Burman, *The Wounded Soldiers of Industry. Industrial Compensation Policy, 1833-1897* (Oxford, 1983). A follow-up study by P.W.J. Bartrip, on the workmen's compensation system between 1897 and 1948, is in preparation.

2   Statistics for non-fatal accidents are notoriously unreliable and can be misleading. John Benson argues, in respect of coal-mining, that the trade unions massively underestimated the extent of non-fatal injury and that at the end of the nineteenth-century more working days were lost through injury than through strikes and lockouts. See J. Benson, 'Note on Non-Fatal Coalmining Accidents', *Society for the Study of Labour History Bulletin*, vol. 32 (1976), pp. 20-22. See also P.E.H. Hair, 'Mortality from Violence in British Coal Mines, 1800-50', *Economic History Review*, 2nd ser. , vol. 21 (1968) , pp. 545-61. Some of these issues are discussed in chapters 1 and 2 of Bartrip and Burman, *The Wounded Soldiers*.

3   *Parliamentary Papers* (hereafter *PP*) 1942-43 VI, Inter-Departmental Commitee on Social Insurance and Allied Services , Report, p. 41.

4   (1837) 3 M. & W. 1; 3 Murph. & H. 305; L.J. 7 Ex. 42; 1 Jur. 987.

5   *Lincolnshire Chronicle and General Advertiser*, 22 July 1836.

6   See: T. Ingman, 'The Origin and Development up to 1899 of the Employer's Duty at Common Law to take Reasonable Care for the Safety of his Employee', unpublished Ph.D. thesis, Council for National Academic Awards, 1972.

7   See Harold Perkin, *The Origins of Modern English Society, 1780-1880* (London, 1969).

8   National Register of Archives, Broadland Mss. , Shaftesbury Diaries, SHA/PD/2: 24 Aug. , 16 Sept. 1840. See *PP* 1841 X, Report of Inspector Howell, pp. 167-69; Bartrip and Burman, *The Wounded Soldiers*, p. 20 and 25-28.

9   See D. Duman, 'The Judges of England 1730-1875: A Social, Economic and Institutional History', unpublished Ph.D. thesis, Johns Hopkins University, 1975; J. Morgan, 'The Judiciary of the Superior Courts, 1820-1968: a Sociological Study', unpublished M.Phil. thesis, University of London, 1974.

10   (1842) 38 Am. Dec. 339; 3 Macq. 316; 4 Met. 49.

11   5 Ex. 343; 6 Ry. & Can. Cas. 580; 19 L.J. Ex. 296; 15 L.T.O.S. 230; 14 Jur. 837.

12   11 and 12 Geo. 6 c. 41. In the half century before 1948 the practicability of a defence of common employment had been significantly eroded by legal decisions, notably in *Groves v. Wimborne* (1898) 2 Q.B. 401, *Lochgelly Iron and Coal Co. v. M'Mullan* (1934). A.C. 1 and *Wilsons & Clyde Coal Co. Ltd. v. English*(1938) A.C. 37, as well as legislation, especially the Factory Act, 1937. See A. Russell-Jones, 'Workmen's Compensation, Common Law Remedies and the Beveridge Report' , *Modern Law Review*, vol. 7 (1944), pp. 19-21.

13   The Law Reform (Contributory Negligence) Act, 1945 (8 & 9 Geo. 6 c. 28) allowed a judge to reduce damages to the extent that he found the claimant to have been responsible for his misfortune.

14   Bartrip and Burman, *The Wounded Soldiers*, p. 76; *PP*1873 XIV, Select Committee on the Regulation of Railways (Prevention of Accidents Bill), Evidence, pp. 587-88.

15   *PP* 1876 IX, 1877 X, Select Committee on Employers Liability for Injuries to their Servants.

16   43 & 44 Vict. c. 42.

17   See Bartrip and Burman, *The Wounded Soldiers* , pp. 158-73; S. & B. Webb, *Industrial Democracy* (London, 1897), pp. 374-75; T. Ingman, 'The Origin and Development', p. 165; J. Benson, 'The Compensation of English Coal Miners and their Dependants for Industrial Accidents, 1860-1897', unpublished Ph.D. thesis, Leeds University, 1974, p. 174.

18   4 *Hansard* 8 (20 Feb. 1893) c. 1961.

19   See *PP*1846 XIII, Select Committee on Railway Labourers, Evidence, pp. 585-89; E. Chadwick, *Papers read Before the Statistical Society of Manchester, on the Demoralization and Injuries Occasioned by the Want of Proper Regulation of Labourers Engaged in the Construction and Working of Railways* (London, 1846), pp. 18-19. See also, R. A. Lewis, 'Edwin Chadwick and the Railway Labourers', *Economic History Review* , 2nd ser., vol. 3 (1950) pp. 107-18.

20   See 'Report of the TUC Parliamentary Committee' in *TUC Proceedings* (1897), p. 22.

21   Workmen's Compensation Act, 1897, 60 & 61 Vict. c. 37.

22   W.C. Mallalieu, 'Joseph Chamberlain and Workmen's

Compensation', *Journal of Economic History* , vol. 10 (1950) p. 57.

23 Throughout this section statistics are drawn from annual Home Office compilations published as Parliamentary Papers.

24 *Balfour Committee on Industry and Trade. Factors in Industrial and Commercial Efficiency. Being Part 1 of a Survey of Industries, Final Report* (HMSO, London, 1929), p. 255.

25 23 June 1928.

26 Public Record Office (PRO) PIN 12/101. Evidence of W.H. Thompson to Departmental Committee on Alternative Remedies, 28 Feb. 1945, p. 32.

27 Nuffield Social Reconstruction Survey (NSRS), Box 202 [1983], dated June 1942, on Workmen's Compensation, p. 5.

28 In the original legislation compensation was to be calculated entirely by reference to previous earnings. The 1923 Amendment Act introduced small supplements in fatal cases where dependants aged less than 15 were involved. This idea was extended during World War II when widows' benefits were introduced.

29 NSRS, Box 202, dated June 1942, on Workmen's Compensation, p. 9.

30 The original Act gave workmen's compensation beneficiaries first claim on the assets of an insolvent employer. In order to exclude insecure companies from undertaking workmen's compensation business, the Employers' Liability Insurance Companies Act, 1907 (7 Edw. VII c. 46) required companies, with some exceptions, to deposit £20,000 with the Board of Trade. Compulsory insurance in the coal trade dated from 1934 (24 & 25 Geo. V c. 23).

31 A. Parsons and A. Bertram, *The Workmen's Compensation Acts 1897 and 1900*, 2nd edn (London, 1902), p. v.

32 *Hoddinott v. Newton, Chambers & Co.* [1901] A.C. 63.

33 A.H. Ruegg, *The Laws Regulating the Relations of Employer and Workmen in England* (London, 1902), p. 151.

34 [1911] A.C. 410.

35 See, for example, Sir A. Wilson and H. Levy, *Workmen's Compensation* (2 vols., London, 1939), vol. 1, p. xiv; G. Howell, *Labour Legislation, Labour Movements and Labour Leaders* (2 vols., London, 1905 edn), 2, p. 432; V.R. Aronson, *The Workmen's Compensation Act 1906* (London, 1909), p. 17; *Derbyshire Times*, 3 Dec. 1898; *The Times* (4 Oct. 1898 and 24 July 1899).

36 'Industrial Law', *Law Quarterly Review* , vol. 51 (Jan. 1935) pp.

197-98; see also his 'Industrial Relations and the State: A Reform of Workmen's Compensation', *Political Quarterly*, vol. 1 (1930), pp. 511-30, and W.A. Robson (ed.), *Social Security* (London, 1943), p. 13.

37   S.H. Noakes, *Butterworths' Digest of Leading Cases on Workmen's Compensation* . . .(London, 1933).

38   See, for example, *PP* 1902, XCVII, Statistics of the Proceedings in County Courts in England and Wales under the Workmen's Compensation Act, 1897 . . . During the Year 1900, p. 621. Others also took this line, see PRO PIN 12/1, Report of His Honour Judge W.C. Smyly to the Home Office, 16 Jan. 1904; *PP* 1904, LXXXVIII, Departmental Committee Appointed to Inquire into the Law Relating to Compensation for Injuries to Workmen, Report, p. 763.

39   According to the Beveridge Report workmen's compensation litigation generated 'legal expenses on a scale exceeding that of the other forms of social security in this country, or of compensation for industrial accident or disease in other countries'. *PP* 1942-3, VI, Report of the Inter-Departmental Committee on Social Insurance and Allied Services, pp. 154-56.

40   *PP* 1932-3, XXVI, Statistics . . . 1931, p. 1019-20; *PP*1933-4, XXVI, Statistics ... 1933, pp. 1271-72. See 5 *Hansard*, 235 (27 Feb. 1930) cc. 2393-4; 268 (6 July 1932) cc. 473-6; 270 (15 Nov. 1932) cc. 953-4; vol. 286 (2 March 1934) c. 1432.

41   Workmen's Compensation (Coal Mines) Act, 1934, 24 & 25 Geo. V c. 23.

42   See, for example, 5 *Hansard* 212 (20 May 1936) cc. 1197-8.

43   *PP* 1920, XXVI, Departmental Committee on Workmen's Compensation, Report, pp. 13-16.

44   PRO PIN 11/7, dated 3 Jan. 1922.

45   *PP* 1923, XIX, Undertaking Given by the AOA on behalf of its Constituent Insurance Offices for the Purpose of Limiting the Charges to Employers in Respect of Employers' Liability Insurance, pp. 555-58; see also *The Times* , 16 June 1923; PRO 11/5, Workmen's Compensation (No. 2) Bill 1923, undated (date stamped 14 Dec. 1923) and unsigned memorandum from the Home Secretary to the Cabinet; *Balfour Committee, Final Report* , p. 256.

46   16 June 1923.

47   'Industrial Law', p. 198; see also Robson, 'Industrial Relations', pp. 511-30; E.H. Downey, *Workmen's Compensation* (New York,

1924), p. 100.

48 Beveridge Papers (British Library of Economic and Political Science), BP VIII 32, 'Further Thoughts on Workmen's Compensation', 17 July 1942. See Jose Harris, *William Beveridge. A Biography* (Oxford, 1977), p. 400. Beveridge's ambivalence about insurance is evident in the Social Insurance Committee's report when he managed to list it as both an advantage and disadvantage of workmen's compensation.

49 *PP* 1920, XVI, Department Committee on Workmen's Compensation, Report, p. 67.

50 *PP* 1904, LXXXVIII, Departmental Committee Appointed to Inquire into the Law Relating to Compensation for Injuries to Workmen, Report, p. 749.

51 *PP* 1942-3, VI, Report of the Inter-Departmental Committee on Social Insurance and Allied Services, pp. 153-54; see BP VIII 32, undated and unsigned memorandum entitled Workmen's Compensation — Present Position.

52 *PP* 1943-4, VIII, Workmen's Compensation. Proposal for an Industrial Injury Scheme, p. 532.

10 WHAT IS AN ACCIDENT?

Karl Figlio

I Introduction

The idea of an accident seems straightforward. It is an unforeseen event which is also expected; in this sense, it conforms to our notions of natural law and fits a pattern. The number and the kinds of accidents show regularities, but the moment of any one accident remains unknown, although it is often retrospectively 'predictable'. In the same way, modern medicine sees illnesses as accident-like events; the specific aetiology of disease treats the overall burden of illness as a regular feature of a society, but interprets any single episode of illness in terms of a pathological injury, say, by germs.

The purpose of my paper is to make this commonsense notion historical. So deeply rooted in our cosmology as an ordinary accident may be, it remains a feature of an historically constituted way of understanding events. The notion of an accident arose along with the dominance of contract-based social relations and the settlement of disputes in those relationships on the basis of contract ideas. In particular, I shall deal with occupational accidents in the context of employment contracts, whether explicit or assumed. An occupational injury or illness was an accident at work, and, therefore, could be formulated as a distinct event only when employment contracts became the dominant model of the social relations of work. That was a feature mainly of the nineteenth century, though clearly it had been developing over a long period.

Most of my material comes from claims for compensation under either common law or the statute law of the Employers' Liability and Workmen's Compensation Acts of the late nineteenth and early twentieth centuries, which sought to clarify and specify certain aspects of common law. Medico-legal adjudication laid down a framework and an everyday practice within which an event could be seen to be accidental; more strongly, it articulated or perhaps even postulated the notion that nature worked in such a way as to produce expected, yet unforeseen events which happened without any apparent cause.

I see this possibility contained within the contract-form; in particular I see the contract-form constituting the notion of an event which 'just happens'. The contract lays down a set of expected events, both by explicit formulation and by implication: it establishes a field of events which are natural, a background in which things occur. This field of natural expectation is often invaded by (retrospectively) predictable but unforeseen events which can be treated routinely in the form of claims for compensation. No fault is assumed, in the sense of malice, yet one party is held accountable, *as if* he or she were responsible. It endorses a notion of causation as a contiguous set of events leading back to a moment when someone was responsible, not by intending to injure but simply by virtue of unfulfilled terms of agreement. Motive is there, but in a neutralised form; accountability without culpability. This field of neutralised intention is the foundation of the belief in 'natural' events; and the things which happen within the purview of contract, but outside the terms of implied agreement — the natural invasions into the field of natural events — are accidents.

II Accidents and the Contract-Form of Employment

The Workmen's Compensation Act of 1897 established a procedure to pay compensation to injured workmen as a routine matter, if they suffered an 'injury arising out of and in the course of employment'. Within strictures on the definition of injury at work, the Act replaced litigation by administration. The workman no longer had to prove employer liability; nor could the workman be deprived of compensation for other than 'serious and wilful misconduct'. The Workmen's Compensation Act of 1906 extended the 1897 Act to a large number of previously excluded occupations, and included a small number of specified diseases which were to be treated 'as if' they were injuries under the 1897 Act. So, by the early 1900s, injuries and some diseases routinely entered employment statistics as unexceptional events to which no liability attached; they simply happened.[1]

Before the Workmen's Compensation Act, the injured workman could still make use of the Employers' Liability Act of 1880, of common law and, if malice were involved, of criminal law. In every instance, the worker had to show that the employer had been responsible for the injury. To prove liability was to demonstrate that the injury did not just happen; it was caused, no matter how indirectly, by the employer.

The Employers' Liability Act simplified the demonstration of employer responsibility, but liability still had to be proven.

Unlike the relationship between two parties of equal status, the relationship with respect to compensation for injury between employer and employee was skewed in the employer's favour. Suits against employers under common law were rare during the nineteenth century.[2] Foster has argued that the employer/employee relationship was employment or service 'at will', which meant that the employer agreed to pay for work just completed, on a minute-to-minute basis. The employer therefore accepted no responsibility for the worker beyond the moment and beyond paying for what he had in hand from the worker.[3] Foster does not include employer liability for injury in his study, but one can derive from it further evidence of the difficulty in establishing contractual agreements on responsibility for injuries at work.

The Workmen's Compensation Act of 1897 put aside this requirement to demonstrate liability. One can analyse the shift from accountability to routine happenings at several levels. Elsewhere in this volume, Peter Bartrip describes the 1897 Act as very much a compromise issue, and he points out that the conversion from litigation to administration hardly did away with court proceedings. It also left the trade unions worried that there remained little incentive for employers to improve safety standards. What concerns me here, however, is the principle made explicit by the Workmen's Compensation Act that an injury could occur which was nobody's responsibility, but which fell to the employer to compensate, because it arose 'out of and the course of employment'. The injury — or the disease — became an accident.

If we put labour law to one side, we see that there was a long tradition of obtaining compensation through civil actions for injury resulting from unintended actions. A compensatable accident could, for example, arise from unexpected consequences of actions agreed between contracting parties. At least as far back as the fourteenth century, common law provided for redress against a doctor who was taken to have breached his contract by not curing a patient. [4] Such a remedy is less surprising in the light of recent research by Margaret Pelling, which shows the inextricable interlocking of medical service and medical trade in the early modern period, so that a material product was often involved in the contract between doctor and patient. [5] The early cases of breach of contract by not curing were actionable as 'negligent misconduct after an undertaking'. It was a borderline area, acting outside the intended range of actions, in such a way as to harm someone of equal status — someone who could be party to a contract, and who could bring a claim for compensation for a breach of the contract.

As I have indicated, an employee (servant) was rarely in the position of an equal party to a contract with an employer (master). Because of this inequality, liability for a servant's actions reached back from servant as instrument to master as actor. A claim could also arise from injury to a servant, but as an action for depriving a master of his employee's services. In these cases, although the servant was the physically injured party, the master remained the legally injured party. The legal relationship and the grounds to bring an action existed between the two masters. In discussing actions arising in the context of master and servant, *Blackstone's Commentaries* for the early nineteenth century refer only to this sort of case.[6]

The concept of an accident was foreign to such a master/servant relationship. Even in relations between equals, when a contracted set of actions, agreed by consenting parties, had an unexpected outcome or ancillary consequences, the claim lay in showing misconduct *(malfeasance)*.[7] *Malfeasance* still presupposed an act, albeit wrongful or negligent; and the legal relationship between super- and subordinate — as in master and servant — included the subordinate's acts within the superordinate's liability. Thus, every act implied responsibility. Society was a space dense with responsibility and liability.

My argument is that the growing prominence of contract relationships and contract law, mainly from the sixteenth century for the general case, and the nineteenth century for master/servant relationships, established the possibility of an accident. During the sixteenth century, the legal formalism for establishing the terms of contract and its breach — *assumpsit* — began to apply to cases of not doing *(nonfeasance)*, as well as to *malfeasance*.[8] The simple failure to fulfil a contract could deprive another person of the opportunity to realise a profit; not doing could therefore do harm, and negligence became a form of action.[9] Accident could only be formulated in law when a breach of contract by *nonfeasance* became possible. It could only exist in a cosmology in which the common view of personal relationships had become contractual, so that obligations and injury could be seen as terms of contract, rather than as motivated acts. Taking occupational accidents in particular, I shall suggest that they could only occur when such relations existed between master and servant. First, I'll sketch the change in cosmology, from a society in which everything had a cause, to a contract society, in which accidents could occur.

III Events as Acts

When we look at injury and illness in traditional cultures, we are not surprised to find that they are symbolically important indicators of social integrity and disruption. [10] That they happen to an individual is less significant than their meaning for the social group. We are less familiar with such symbolism in modern western culture. Recent work on community structure in Germany, however, shows a continuous surveillance of social indebtedness and honour among village families.[11]When something occurs in such a setting, either fortune or misfortune, it is expected. Something like that was about to happen and not necessarily to anyone in particular, though 'just deserts' might frequently be commented upon. An injury, an illness, or some other misfortune, brings relief from the rising community temperature. Nothing could happen by accident, when every event is scrutinized for its place in the dense fabric of expectation.

Legal relations are, of course, more formalised than unwritten community expectation. When something goes wrong, legal procedures make a judgment according to codified procedures. But the procedures build upon common understanding of events, and they allow us to glimpse the social construction of fundamental categories of thought.

Coroners' inquests, for example, have been central to the development of a concept of accident. [12] The coroner had to establish whether or not a death resulted from malice or from a misfortune to which no intent could be attached. Well into the nineteenth century, coroners' verdicts on occurrences that we would call simply accidental death show a curious mixture of act and accident. Their verdict of 'deodand' singled out from the many unexceptional but sudden deaths a particular kind of event, described in the following way around 1800[13] :

> If a horse, or ox, or other animal, of its own motion, kill as well an infant as an adult, or if a cart run over him, they shall in either case be forfeited as deodands; which is grounded upon this additional reason, that such misfortunes are in part owing to the

negligence of the owner, and therefore he is properly
punished by such forfeiture.

Examples from a series of nineteenth century verdicts include: trodden
by horse, fell from gig, kicked by horse, hit by falling timber. A
substantial portion of accidental deaths were classed as deodands. [14]
The forfeiture of the object 'moving to the death' of the victim —
perhaps only a cart-wheel, but sometimes the cart, its contents and the
horses — were valued, and the proceeds were distributed among the
poor, or given to the lord of the manor, the crown's agents, or the
church.

Deodand in the routine coroners' proceedings revealed an archaic
cosmology, like the village world which I described earlier, contained
within modern rational social practices. It suggested an accidental, yet
curiously intended, occurrence, so that the wrong-doing had to be
atoned for, even though the instrument was an animal or an inanimate
object. Indeed, deodand verdicts sometimes included explicitly the
notion that the instrument of death had acted feloniously; and, as we
shall see later, the legal formalism for felony did carry over into liability
from breach of contract. [15]

Coroners' decisions rested on the usual legal refinement of evidence
and argument in relation to precedent, and the records of inquests
show a well-established routine procedure. For example, one series of
inquest records had been sent on to another court for deposit, with a
covering note suggesting that the rulings should have been deodand,
and not simply accident.[16] An eighteenth-century coroners' guide
recounted several versions of an apparently often-used precedent for
deodand, in which a man drowned in a violent stream while crossing
with horses. The author was not convinced that deodand had been the
verdict, and therefore doubted that this case could be used as a
precedent.[17]

These routine legal proceedings, with their discrimination of details
in the search for the immediate agent, rested ultimately on a belief that
things did not just happen. Deodand might therefore be seen as a
transitional notion, a category of ambiguous accident, unforeseen and
not malicious, yet somehow implying intent. It is reminiscent of the
village, in which social debts had accumulated and were being
discharged through a misfortune.

The quotation from Blackstone points up another strand in
deodand, again of a transitional sort. Deodand justified punishment
on grounds of negligence, as opposed to wrong-doing. One can discern
in the category of deodand a shift in cosmology from one in which

everything was caused, so that cause in the physical world was equivalent to motive in the human world, to one in which some things just happened. Deodand included both.

With negligence as a cause of an accident, the key feature of the modern contract-form of social relations emerged. *Nonfeasance* could harm, and lead to legal action, just as *malfeasance* was actionable. As an occurrence stripped of intention and wrong-doing, an accident was an event lying outside the explicit and implicit agreements between consenting parties to the contract; it was like a breach of contract. Put in an extreme and stark form, an occupational accident became possible as the relationship between employer and employee became contractual.

IV Liability in the Master/Servant Relationship

Contractual relations between master and servant did not commonly exist, even well into the nineteenth century; an employee, therefore, did not have the same rights to redress against his or her master as did a stranger. The crucial case was *Priestley vs. Fowler* (1836), [18] in which an overloaded van, driven by a fellow worker, injured an employee. The court would not award compensation. With no precedent available, it turned to the general principle that liability must be limited, in order to avoid an extension beyond the employer to, say, the coach-maker. From this ruling grew the doctrine of 'common employment', according to which an employer remained free of liability for injury caused by negligence of a fellow worker, even if the latter were a manager. Two points stand out: firstly, the claim failed because an employee did not have the same contractual standing in relation to the employer as a stranger would have had ;[19] secondly, the decision lay within the general thrust of contract law, to limit liability and to emphasize consent between the parties. [20] The contract *form* is visible, even in the limitation of contractual equality in the particular case of labour relations.

Employment was a form of hiring; if the worker did not properly assess the risks that he took on as part of his contract, then he himself breached the contract by not possessing the skills which he offered for hire.[21] Workers did not have to take unreasonable risks, so, in undertaking a job (as in *Priestley vs. Fowler*), he or she took on the associated risks and the responsibility to judge them. [22] The case of

Fowler vs. Lock (1874) turned on whether the relationship between the parties was master/servant or bailor/bailee. [23] The plaintiff, a cab driver, had hired a cab and horse from the defendant. The horse was fresh from the country, with no experience of pulling cabs, and overturned it, injuring the driver. In the original trial, the court awarded the plaintiff fifty pounds, having found their relationship to be bailor/bailee. The defendant then sought to reverse the ruling, arguing that their relationship was master/servant. Although other issues entered into the various appeals, the final verdict affirmed the original decision, mainly by dismissing the appellant's claim to a master/servant relationship with its implication that the cab driver had taken on the risk of the horse's fitness.

The limited liability of master for servant was part of their instrumental relationship, but it also followed from the growing contract form of relationship. The liberal market ideology expressed the conviction that risks were accepted in return for wages, and that wage rates reflected the degree of risk. [24] Adam Smith explained wage differentials on these grounds; in occupations with risks which could not be met with 'courage and address', but were simply unwholesome, wage rates were 'always remarkably high'. [25] Indeed, the implied contract of labour with wages in consideration of risk, could even extend to work in which the risk from negligence of other employees was 'so much a natural and necessary consequence of the employment. . . that it must be included in the risks which are to be considered in his wages'.[26] This last formulation might be seen as a contract formulation of the doctrine of common employment which, in *Priestley vs. Fowler,* was argued from general principles.

Just how limited was the liability of master to servant for injuries sustained at work can be appreciated from the statements of the Lord Justice of Appeal, William Baliol Brett, giving evidence before the committee on the Employers' Liability Bill in 1876. [27] The committee asked him to comment on liability arising under four headings: (1) intrinsic danger of employment; no remedy for servant (2) defect in plant or machinery; master not liable, unless the defect was both known to the master and unknown to the servant (3) negligence or want of skill of fellow workman; master not liable (4) negligence of workman; master not liable.

The Employers' Liability Act (1880) did make two statutory changes in the workman's position at common law. Firstly, it put the workman on the same footing as a stranger, with respect to rights to compensation for injury by a fellow workman; i.e., it abolished the doctrine of common employment. Secondly, the employer no longer had to know of a defect in plant or machinery, in order to be held liable for injury to a workman; he had only to be found negligent in continuing to use it. But equally important, the employer could be liable for not doing something, i.e. for *nonfeasance*. With this latter change, we see the essence of the contractual relationship, which established the possibility of an accident.[28]

V Negligence and the Implied Contract

Negligence lies at the heart of contractual, as opposed to personal, relationships. In the latter, an act of one person may harm another; in the former, failure to act may harm the other. Breach of contract through negligence came into common law in the sixteenth century. In effect, negligence came into master/servant law only during the nineteenth century.

A promise to act according to an agreed intention establishes the contract, whether or not all the implications of agreement are clear.[29] The grounds of a breach of contract might, therefore, not be obvious. Legal proceedings elicit and trace out from the mass of evidence both the intent which grounds the contract and the consequences which follow 'naturally' from breaching the contract.

Let us look more closely at the consequences which follow 'naturally', and therefore become implied terms of contract under common law. In the following case from 1854:[30]

1) A flour mill sent a broken mill shaft to the manufacturer by common carrier.

2) The common carrier delivered the shaft late, thereby breaching their contract with the mill.

3) The mill had no spare shaft, so that it had to close down; but the common carrier did not know there was no spare.

The court allowed breach of contract and compensation, but not the suit for loss of potential profits. It acknowledged a loss, but ruled that the damaged opportunity for profit did not 'arise naturally' from the delayed delivery and that the special circumstance of the mill's closure was not communicated to the carrier.

In this example, we see an interpretation of a contract framing the notions of what is promised in the contract and what events cohere 'naturally', so that they are embedded in the relationship between the parties, even if only by implication. Such a determination was especially difficult in the case of injury or illness, where establishing the consequences 'as would in the ordinary course of things naturally arise from the breach of contract' was complicated by the continually changing state of a person's health, including the effects of his or her own actions. When a railway company conveyed someone to the wrong place, the court allowed compensation for the inconvenience, but not for the illness following walking home in bad weather or for the attendant medical expenses. 'The latter head of damage claimed was disallowed, as being merely connected with a breach of duty by "a series of causes intervening between the immediate consequence of the breach of duty and the damage complained of" '.[31]

The restrictions on employees were more severe. Unlike the relationship between bailor and bailee or seller and buyer, the master/servant relationship implied the continuing acceptance of risk. Although a buyer might claim compensation under an assumed warranty for misrepresented goods, a worker was held to have assessed risks of all sorts, whether of plant, of machinery, of the labour process or of other workers. Whatever the restrictions on liability already contained in the traditions of the master/servant relationship, the contract-form of employment in the nineteenth century formalised the recognition and acceptance of risk as a natural feature of employment (*volenti* principle). [32]

For example, although the Employers' Liability Act explicitly did away with the doctrine of common employment, it left the contract form, with the *volenti* principle, quite intact. And the contract form included risks arising from actions of fellow workers. The following passage from Addison's classic text on contracts continued to appear even in 1911, long after the implementation of both the Employers' Liability and Workmen's Compensation Acts: [33]

> [A] servant when he engages to serve a master, undertakes as between him and his master, to run all the ordinary risks of service, including the risk of negligence upon the part of a fellow servant, when he is acting in 'the discharge of his duty as servant of him who is the common master of both'.

Thus a worker's injury might inhere in the job, always there as a possibility within the contract of employment. Legal and medical scrutiny focused on the implied contract, in order to discriminate events which formed a natural cluster within the contract from those that did not. The implied contract and its unintended breach constituted a field of unexceptional occurrences which, therefore, required no particular explanation. In a way which the pre-contract cosmology did not allow, these events just happened. The implied contract and its unintended breach mutually established the notion of 'by nature' and 'by accident'. I shall develop this point further with two examples. Both arose under the Workmen's Compensation Act, but this source of litigation doesn't affect my argument about accidents. Indeed, the Workmen's Compensation Act was built on the principle of accidental injury and accident-like disease.

In the first case (1909), a London sewer worker contracted enteritis from sewer gas, which accelerated a long-standing heart disease and consequent incapacity. This progression of events was accepted by the county court judge, but the judge none the less ruled against the workman, because enteritis was a risk incident or intrinsic to his job and not an accident. The decision was upheld on appeal. The judge also underlined the interpretation of disease as accident, by pointing out that the legislation to extend coverage of the Act to scheduled diseases treated these few diseases by statute *as if* they were accidents arising out of and in the course of employment; it did not, therefore, say that *they* were accidents. The workman's enteritis was neither scheduled, consequently to be accepted as if it were an accident, nor did it breach his contract. Incident to his employment, enteritis fell within an implied contract, according to which master and servant shared the intent to exchange risk for money. [34]

In the second case (1909), a collier fractured his shoulder at work. The surgeon splinted it and gave him directions for arm exercises. He received compensation, and after two and four months, other surgeons examined him again and gave him further instructions. Eight months after his incapacitation, the collier's shoulder was still stiff. Meanwhile, the company stopped his compensation, on evidence from the surgeons that his continued incapacity resulted from his failure to follow instructions.

The court established that the collier was 'of very nervous temperament, and that this constitutional and natural nervousness, intensified to some extent by the accident, was the reason why the applicant did not carry out the directions of the medical men; that this neglect had delayed his recovery, and was the cause of his present inability to work; and that this neglect to obey the instructions was not the result of wilfulness or carelessness, but was due to the nervous condition which he appeared unable to control.' The judge reinstated his compensation, and this decision was upheld on appeal. [35] In this case, as in the first, there was a natural clustering of events; here both physical injury and psychological state combined to cause incapacity. The difference between them lay in the relationship of the naturally occurring events to the implied contract. In the former case, they were implied in the contract; in the latter case, they were not. They were accidents (of the contract).

The idea of 'naturally occurring' comes out of normal common-law thinking about both torts and contracts. It requires further elaboration, and I will come back to it. First, let us look in more detail at the demarcation between events implied in a contract and events not implied in a contract. The distinction lay in whether or not a discrete incident could be identified, which happened within the overall purview of the contract, but not within any explicit or implicit acknowledgement of such a possible occurrence by any party to the contract. The pre-existent condition of a worker or of equipment mattered less than any changes in condition within the purview of the contract.

A comparison between *Fowler vs. Lock* (the case of the cab-and-horse hiring) and *Robertson vs. Amazon Tug and Lighterage Co.* will illustrate this distinction between conditions which breach a contract and those which do not. [36] In *Fowler*, the plaintiff had hired equipment for a specified period and a specified job. The bailor implicitly agreed to provide equipment appropriate and adequate for the contracted job, and his failure left him liable to compensate the bailee for his injury. In *Robertson*, a master mariner hired a particular ship. The ship's boilers needed continual attention, which delayed his voyage. He based his claim for compensation on *Fowler*, but lost, because the *particular* equipment that he hired had not further deteriorated during the period of the contract. The ship's condition was unexceptional, and no accident occurred. The bailor in *Fowler* negligently breached his contract; the bailor in *Robertson* did not.

In the collier's case above, he did not abrogate his contract by his nervousness (he had worked), but the colliery did negligently breach the contract (in a statutory way) when the original injury occurred. A similar interpretation applied to a man who left his home, apparently in good health. His work that night was heavier than usual, but no heavier than at some other times. Shortly after work, he collapsed, and later died. Post-mortem examination showed disease of the heart. With no evidence of a particular strain before him, the judge ruled that death resulted from a primary and pre-existent (idiopathic) illness, not from an occupational cause. On appeal (1938), his decision was reversed, with the conclusion that the workman's death had been accelerated by his normal work; and his widow's claim was upheld. [37] A similar decision was taken, following a death from an aneurysm while a workman was tightening a nut (1910).[38] This sort of case extended readily to include disease by poisoning or infection, so that when a man employed in handling manure consisting mainly of bone dust suffered fatal blood poisoning, the claim succeeded. [39] So too did the claim which set a precedent for the scheduling of occupational diseases under the Workmen's Compensation Act of 1906. In this case (1904), a wool sorter contracted anthrax which, in the court's formulation, was the consequence of a stray germ present in the wool, settling on a sensitive part of his body, such as the eye or a sore. [40]

Sometimes, no injury could be found. A labourer doing very heavy work received compensation for muscular strain to his chest. He later sought a memorandum of agreement to compensation based on the initial award. The court rejected his claim, using the following arguments: firstly, his muscular strain was cured; secondly, he had been suffering from myocardial heart disease throughout the period; thirdly, every sort of work strained him severely and promoted the disease and, fourthly, the disease was progressive by nature, its rate accelerated in proportion to exertion. [41]

If a supervening event at work, within the purview of contract, occurred, then a new chain of events — a new natural history of injury or disease — began from that moment. The new event entailed a new set of sequelae, all legitimated for compensation by the legitimacy of the new event, which was unimplied, but within the purview of the contract. In this case, an accident occurred, according to the principle of the *novus actus interveniens*.

My analysis might suggest that the test to determine whether or not an accident occurred was straightforward. It was certainly easy to state, but more difficult to apply, or to understand the grounds of a particular application. Consider the difference between the following two cases. In the first example, a plate layer had been working in a railway tunnel over a long period, during which time a full railway service operated. His claim, on grounds of anthracosis, a dust disease, presumably from locomotive smoke, was rejected. [42] In the second case, a girl suffered poisoning from the effects of several cuts sustained at work. Although no single accident caused her incapacity, her claim succeeded. [43]

The distinction was drawn between the successive impingements of dust, in the first case, and the cumulative effect of accidental cuts, in the second. The first case illustrates a natural history of disease incident to the job, a situation in which the successive exposures to coal dust simply repeated or maintained the background conditions of the job. The second case illustrates an accident with a cumulative impact. Presumably each cut stood out from the background conditions of the job, so that their aggregated effect could be seen as an accident, even though no single cut had any effect and the repetition of minor cuts might have seemed almost normal for the job. None the less, the risk from smoke lay within the implied contract; the cuts did not.

VI Nature, Facts and the Contract

We can see from the examples in the previous section that injury and disease were brought into the same framework. The same test of a breach of the implied contract applied both to a fracture and to coronary heart disease. And none of the statutory regulation of liability over-rode the fundamental feature of contract, the ascertainment of agreement with its implied terms; the hallmark of the labour contract remained the so-called '*volenti*' *(non fit injuria)* clause.

The *volenti* clause meant that the worker accepted the risks inherent in a job as implied terms of contract, by the very fact that labouring for money was a voluntary act — a contract. Such a logic is reminiscent of Marx's ironic use of the term 'free labour' to describe the freedom of a peasantry — converted to a proletariat — to take up work. Yet, free or driven, the worker within a contract-form of employment, entered a working relationship in which the terms of working were carried along in the same momentum towards explicit clarification as they were in other areas of contractual relationships. One judge summed up the *volenti* principle: [44]

> The principle embodied in the maxim ... has sometimes been stated thus:- A person who is engaged to perform a dangerous operation takes upon himself the risks inherent thereto. To the proposition thus stated there is no difficulty in giving an assent, provided that what is meant by engaging to perform a dangerous operation, and by the risks thereto, be properly defined.

In other words, provided a contract genuinely existed — provided all the items, explicit and implicit, had consent — then the proposition was acceptable. The quotation above shows the continual clarification of the implied contract, in that the nature of *volenti* was more carefully defined; agreeing to work could not be taken in itself as evidence of agreement to assume all risks of a job.

The contract therefore became more the articulation of an established consensus than a formalism to impose regulation. Of course, for most working people, consensus must have been as ironic a notion as was free labour. None the less the formulation of a relationship as contractual, then of the notion of contractual as consensual, brought a hidden implication clearly to the surface: if the parties agreed to the explicit and implicit terms of their relationship, they had by definition made a contract; if they had not agreed, then they had not made a contract. The contract either existed or it did not. The existence of a contract was a matter of *fact* .

To say the contract was a matter of fact was also to say that it was not a matter of law. The finding of fact displaced the judicial interpretation of the events. The Court of Appeal clarified the factual vs. the interpretive mode of establishing the right to compensation under the Workmen' s Compensation Act in the case of a miner who injured his knee at work (1910). The company had paid compensation for ten months, when they applied for review and termination of payments. Meanwhile the miner had developed asthma which had become the cause of his incapacity. His original injury had occurred on a cold day. During the two hours it took him to get home, he had suffered a chill, followed by chest trouble and pneumonia, and ultimately asthma. In response to the company's application, the county court reduced compensation payments to 1d per month, arguing that the miner's condition was not the *natural* result of his injury. The Court of Appeal, however, said that it was not for the law to judge the natural or probable outcome of an accident, in assessing legitimacy of a claim based on an ensuing disease. The issue was what actually did occur. In this case, what occurred was an accident, followed by debilitation, followed by disease, followed by the incapacity currently before the court. The Court of Appeal redirected the County Court to apply the proper test. [45]

The classic case which elaborated the notion of an unbroken causal chain was *Dunham vs. Clare* (1902). A factory worker dropped heavy pipes and injured his toe. Over the following two weeks, he attended a hospital outpatient clinic for treatment, during which time he contracted erysipelas and died. Medical evidence was brought in by the employers to show that he must have infected himself some time after his injury, so that his death could not be considered an accident of employment; his death, they argued, was not the natural or probable consequence of his injury. The judge laid down the principle that: [46]

> The question whether death resulted from injury resolves itself into an inquiry into the chain of causation. If the chain of causation is broken by a *novus actus interveniens*, so that the old cause goes and a new one is substituted for it, that is a new act which gives a fresh origin to the after-consequences.

If the chain was unbroken, then the question of natural or probable consequences does not arise.

The Workmen's Compensation Act specified that an injury occurring at work qualified for compensation and automatically became the first link in a causal chain. That specification simply did away with the requirement under common law, that the implied terms of contract be elucidated, in order to discover if the injuring event were a breach of contract. The same principle, therefore, operated in both cases, with the Act simplifying by statute the exploration of the terms of contract, once it was established that the accident occurred at work.

None the less, there was some discussion in *Dunham vs. Clare*, as to whether or not the workman caused his own infection by unwise action, i.e., whether or not he could be held responsible (The case of the collier was similar; ref. 35). In the end the decision based on the unbroken chain rested on an interpretation of the Workmen's Compensation Act. The discussion of the worker's actions, however, shows the traces of contract thinking which, in turn, overlaps notions of tort or wrong-doing. All three legal areas show the attempt to delineate what follows on naturally from an action, either intended or attributable.

The residual sorting out of the distinctions between tort and contract in the early twentieth century highlights the ambiguity which the case of deodand illustrated for the early modern period. Addison on torts clarified the two areas this way: [47]

> The general rule of law is, that whoever does an illegal or wrongful act is answerable for all the consequences that ensue in the ordinary and natural course of events, though those consequences be immediately and directly brought about by the intervening agency of others, provided the intervening agents were set in motion by the primary wrong-doer, or provided their acts were the

> necessary or legal and natural consequence of the
> original wrongful act.

In the case of contract, the implicit agreements of the contract would specify the natural consequences; in the case of employment statute law, the confirmation that an act occurred at work sets in motion a causal chain. In all three cases, there is an actual or stipulated responsible agent. Thus, in many cases of injury, the immediate agent or responsible party to the injury was by-passed, and the claim for compensation extended back to the original agent, as the bearer of overall responsibility.

So it was, that a man who let water run down the street, not realising that the sewer was blocked, was not held responsible for the injury to a horse who slipped on the accumulated frozen water around the corner;[48] but local commissioners whose defective sluice gates flooded properties and, due to defensive reactions by the property owners, flooded the plaintiff's property, were held responsible.[49] The stopped drain was the original offending act in the first case, and the flooding water in the second case. In *Dunham vs. Clare* and in cases in general, where the actions taken by the workmen were set in motion by the original injury, their enchained actions remained part of the natural cluster of events, whether or not they appeared to follow on from the initial injury, and the initially responsible person remained responsible for all that followed. In this sense, the judges working within the Workmen's Compensation Act were within the tradition of contract law, when they refused to allow the question of natural consequences. The workmen's actions taken in response to injury were within the umbrella of responsibility within the implied contract.

The principle of Workmen's Compensation, therefore, was the same as the grounding of contract in consensus. The consensus and its consequences became matters of fact, not of interpretation. That was the principle, and it was consistent with the two aspects of the trial; the judge dealt with matters of law, while the jury dealt with matters of fact. The appeal in the miner's case above (see ref. 35) rested on a mistaken point of law (the county court judge's opinion on the natural outcome of injury), so that the determination of fact had been over-ridden. In an earlier example, the case of the hiring of a cab and horse, the appeal went first to a 'court of errors', whose job it was to establish the existence and nature of the contract between the two parties. Although later clarification was still necessary, the principle remained that the fact of the contract contained the resolution of the dispute.

Natural clusters of events — events which were causally related —

reappeared as contiguous events, either expected or not expected in the contract. A chain of causality which would explain a phenomenon in terms of causal antecedents was reconstituted as a chain of contiguity — a sequence of things occurring within a specified geographical field of application, such as a location. Explanation became an accounting of what did happen, not of hidden causes or of judgements about the nature of things. The idea of getting an account of the facts, from which a resolution would emerge, suggests a legal positivism, corresponding broadly to a positivist epistemology. [50]

The mass of legal precedents and unending attempts to resolve disputes testify to the complexity of human affairs, but they also show a general movement towards a view of nature as aggregates of events. The emergence of a contract-form of human relationships and especially the notion of *nonfeasance* which has been central to it, have helped to form such a view of nature. In other words, the medico-legal machinery represents a massive and cumulative construction of practising with a perception of nature, hidden within the continual practical attempts to resolve disputes.

VII The Accident Model of Disease and Injury

The facticity embodied in the contract-form, with its legal framework for interpreting events in terms of an ultimate, responsible act, setting in motion a chain of 'naturally arising' sequelae, polarised the world-view into two domains: wilful acts and natural facts. The fact of consensus in both explicit and implicit terms of contract drove forward the sense of facticity, and informed a perception of the world as a field of joint endeavour, in which intended actions took place against a background of indifferent events. The social and natural worlds, together having been dense with intention and motive, split into a limited arena of 'voluntary' actions in society and a field of events stripped of intention in nature; now there were foci of contractual relationships enacted against a background of phenomena from which public scrutiny for meaning (as seen in the village example) had been withdrawn. Outside the foci of contract, in the background as the field itself, things could just happen. An accident could occur. [51]

I have argued that the contract-form, and especially the notion of *nonfeasance*, with its implication that not doing anything could be an action, helped to form such a cosmology. Nature as an aggregate of events, rather than of intentions, fits with contractual social relations in which eliciting the implicit agreement in a joint venture turned on discriminating sets of contiguous events which stood inside the contract from those which stood outside the contract. In either case, we are talking of 'facts'. If a venture went wrong for one of the parties, the crucial question was whether or not there existed an event within the purview of the contract, but not implied by it, from which a series of contiguous occurrences followed, and led to the damage suffered. Breach of contract by *nonfeasance* established a field of neutralized intention; in place of actions came events which just happened.

In this cosmology, diseases were also accidents. They were the outcome of a contiguous series of events, set in motion at a discrete time and place. Within the accident mentality, illnesses signified breached contracts. Occupations exposed workers to discrete hazards which could set off a series of contiguously related events. The job was one site at which a contract was honoured or breached; social life was another, and personal life yet another. In this cosmology, life was the aggregate of sites at which discrete processes took place, like exposure to chemicals or to cigarettes or even to inactivity and saturated fats.

The Workmen's Compensation Act was rooted in such an accident mentality, even in the case of scheduled occupational diseases. The schedule lists diseases, considered *as if* they were accidents of employment, i.e. as conditions for which statute put aside the normal elucidation of the implied contract. They were not to be seen as precedents for treating other diseases as occupationally caused. By administrative decree, they were accidents, in spite of being accepted as part of (the nature of) the job. The schedule was divided into two parts; for each disease listed in the left column there corresponded a work process in the right column. Each disease, therefore, was tied to a specific situation; the occupational disease was an event constituted in the correlation of two happenings. At one moment, a pathogenic occurrence set a pathological process in motion.

The Dangerous Trades Committee, set up by the Factory Act of 1895, had already reported on several specific hazards, and required notification of diseases scheduled under the Act. Indeed, four of the six diseases originally scheduled by the Workmen's Compensation Act of 1906 had already been scheduled by the Dangerous Trades Committee. Phosphorus, arsenic and mercury poisoning fit easily into the model of an accident-event; and the fourth (anthrax), unambiguously a disease, was construed as an accident (a 'poisoning' by infectious germs) within normal common-law litigation.[52]

The procedure for scheduling a disease under the Workmen's Compensation Act is clear in the following example. Shortly after the original six diseases were scheduled, the committee on compensation added telegraphists' cramp, and turned down 'twisters' cramp' (Dupuytren's contraction). Telegraphists' cramp was listed with little controversy. Testimony supported the characterisation of a very particular condition which developed from hand movements specific to using the telegraph key. There was no question of confusing it with fatigue or with neurasthenia, because it could be detected, not by a general wearing down, but by certain forms of errors in code sending. Telegraphists' cramp, therefore, was clinically specific, differentially diagnosable and attributable to the specific work process listed on the schedule. [53]

Twisters' cramp in lace makers was not listed until 1921. Although there was plenty of evidence that lace makers suffered from incapacitation caused by the rigid contraction of their fingers against the palms of their hands, there was also evidence that women in the same area, but not employed in lace making, suffered the same incapacity. Twisters' cramp, therefore, could not be attributed only to the specific work process for which it would have been scheduled. Here was an early example of epidemiological specification of occupational illness, such as one finds routinely in later work. [54]

Whether as a scheduled disease, or as a disease compensated, but not scheduled under the Act, or as a disease compensated under common law as a breach of implied contract, the relationship of a disease to occupation was extrinsic; the legal framework constituted disease processes as sets of 'facts' linked by their sequential occurrence. Gill Burke's paper illustrated this way of thinking, in the dominance of the physical theory of phthisis over a more subtle theory of tissue pathology.

This restricted view of nature as an aggregate of accidents, however, also brought together heretofore unrelatable events, including many occupational illnesses, now seen as matters of fact and made legitimate by the contract form. The same contract form which confined the sense of nature to that of contiguity, constituted the notion of accident and threw phenomena into causal relationships that did not previously exist. Countless decisions under Workmen's Compensation linked together illnesses which would not otherwise have been seen to be 'naturally' related.

The accident model of illness also encouraged legislation to control dangerous conditions in factories. It forced home the recognition that production processes were hazardous, in ways which could be analysed and corrected, under threat of penalty. Thomas Arlidge, chairman of the Association of Factory Medical Officers and later Chief Factory Inspector, complained of the failure of doctors to associate specific diseases with specific working conditions. Writing in 1892, he said: [55]

> Attention has been directed almost exclusively to those diseases described in medical nosologies, and to the ordinary catalogue of causes, consequently the lesions and sicknesses attending upon employment have been overlooked, or, if recognised, their pathology has not been followed up...
>
> In registering cases of sickness and death of patients suffering from the lesions [of lung tissue, arising from textile materials] it appears to be deemed sufficient to enter them as examples of bronchitis, or asthma, or consumption, without further investigation.

For Arlidge, an epidemiology which connected pathology with specific working conditions would advance medical knowledge and practice in an essential way.

Along with his call for an aetiology, and not just for a nosological description of disease, Arlidge interpreted the relationship between occupation and illness in contract terms. The effect of his contract-formulation of work and illness was to undercut the epidemiology that he proposed. In typical contract terms, the choice of trade, for Arlidge, was voluntary, and higher wages were an inducement for a worker to take on hazardous jobs. The higher wages attracted chiefly 'the reckless, broken-down characters found in the lower strata of society', so that a recklessness of conduct in life and in health matched the hazards of employment. Even where the occupation wasn't particularly injurious to health, it became harmful indirectly, through its associated moral and social environment. [56]

Arlidge offered up a liberal market ideology applied to health. This is a line which we have followed from Adam Smith to the contract model; one which informed the accident view of illness, both in medicine and in medico-legal jurisdictions, such as employers' liability and workmen's compensation. In the end the worker brought along his social and moral character as background features of his work. They became part of nature, along with the risks which were accepted as intrinsic to the labour process. Both natures were written into the implied contract, leaving occupational illness as the accident that suddenly breached the contract. The illnesses of occupations reflected, in part, occupational hazard but also the nature of the social groups that commonly worked the jobs in question. The distribution of illness in society, therefore, reflected the contracting of employer and employee, and all other implied contracts of human relations, whose adjudication came within common law.

In some ways, the accident view of illness, and Arlidge's use of it, come close to modern notions. Occupation as the index of social class, also indexes all those features of class-based life which have health consequences. Not just exposure to work-process hazards, but also exposure to class-related geographical hazards, such as environmental pollution, poorer access to health care and generally less healthy conditions, contribute to the class dispersion in ill-health. So do the differences in life chances and the cumulative aggregate of harmful life events. All these aspects of epidemiology, whether of work itself, or of the work-indexed life chances and life events, add up to an aggregate of accident-like events which provides a kind of print-out of the health data of a person's life. [57]

What the print-out does not portray is health within a life as it has been lived, including the illnesses, forming a single story. The accident model has produced the data, but the person appears only as the ground on which the aggregate of health-data are displayed. [58] The things that happen outside the contract of the moment belong to another contract. A life as an aggregate of contracts loses its biographical continuity, so that it becomes difficult to see an illness as a present state that emerged gradually from the whole of the past. The accident mentality divides each illness experience from every other, in the quest to establish the moment and the setting which would ground (or abrogate) its legitimacy.

VII Summary

What is an accident? At a specifiable moment, an event led to injury or disease; it was an accident — an unforeseen and unintended breach in natural, social and personal history. If it happened at work, it probably lay within the field of the labour contract, and the question arose as to whether or not the risk was an implied term of the contract. The contract-form of social relations put *nonfeasance* on a par with *malfeasance,* so that doing or intending nothing became a kind of causality, a kind of action. Elucidating the terms of contract determined accountability for injury and illness, and a discovered breach of contract required compensation. The countless instances of legal adjudication of the terms of contract enacted a cosmology within which individual intentions here incorporated as events into lawlike formalised social (contractual) relations. These events could be specified as matters of fact against a background or field in which things could just happen. Even a loss of agreement became a kind of error. In such a field of neutralised intentions, in which things happened by not-doing and appeared as errors among established facts, unforeseen, yet unexceptional, events could occur — as if by nature. Such an event was an accident — a breach of contract.

Acknowledgements : I would like to thank Jeremy Green and Ludmilla Jordanova for discussion and criticism, and Michael Lane and Tony Wells for transcribing and translating early coroners' records.

Notes

1   The classic treatment of workmen's compensation is A. Wilson and H. Levy, *Workmen's Compensation* (2 vols., Oxford, 1939). On the political evolution of the Act, see P. Bartrip and S. Burman, *The Wounded Soldiers of Industry: Industrial Compensation Policy, 1883-1897* (Oxford, 1983). On medico-legal procedures under the Act, see K. Figlio, 'How Does Illness Mediate Social Relations? Workmen's Compensation and Medico-Legal Practices, 1890-1940' in P. Wright and A. Treacher (eds.), *The Problem of Medical Knowledge: Examining the Social Construction of Medicine* (Edinburgh, 1982), pp. 174-224.

2   Proof of employer liability was difficult, even in the cases of violent attack by masters, see T. Forbes, 'Crowner's Quest'. *Transactions of the American Philosophical Society*, vol. 68, no. 1 (1978), see pp. 35-6.

3   K. Foster, 'The Legal Form of Work in the Nineteenth Century: the Myth of Contract?', paper presented to the Conference 'The History of Law, Labour and Crime', University of Warwick,

Sept. 1983.

4   See the Historical Introduction by A. Simpson to Cheshire and Fifoot's, *Law of Contract,* 9th edn by M. Furmston (London, 1976).

5   M. Pelling, 'Medical Practice in the Early Modern Period: Trade or Profession?', *Bulletin of the Society for the Social History of Medicine,* no. 32 (1983), pp. 27-30.

6   *Blackstone's Commentaries on the Laws of England* with notes and additions by Edward Austien, 14th edn (2 vols., London, 1803).

7   There is no entry under 'accident' in Sir Thomas Tomlin, *The Law Dictionary* (2 vols., London, 1820). E. Umfreeville, *Lex Coronatoria* (London, 1761), pp. 450ff. does have a section 'Accidents, Casualties, etc. as well on Land as by Water'.

8   E. Jenks, *A Short History of English Law,* 6th edn (London, 1949), pp. 319-20.

9   Simpson, 'Historical Introduction'.

10   J. Comaroff, 'Medicine: Symbol and Ideology', in P. Wright and A. Treacher (eds.), *The Problem of Medical Knowledge,* pp. 49-68.

11   G. Wilke, 'The Importance of the Perception of Health and Illness for Social Order and Social Conflict', paper presented to the workshop on 'Perception in History and Anthropology: the Problem of "Otherness"', Sponsored by the German Historical Institute, Cumberland Lodge, Windsor Great Park, 22-24 Feb, 1984.

12   On coroners, see T. Forbes, *Crowner's Quest.*

13   *Blackstone's Commentaries,* vol. 1, pp. 300-1; T. Forbes, *Crowner's Quest*, p. 7; W. Wescott, 'A Note Upon Deodands', *Transactions of the Medico-Legal Society of London,* vol. 10 (1910), pp. 91-7. *Coroner's Guide . . . Containing Precedents and Instructions . . .* , 2nd edn (1756). The copy at Cambridge University Library, owned by a local coroner, Charles Martindale, includes manuscript additions, dealing with specific cases from around 1780.

14   Essex County Record Office (ERO), CR/Wl.

15   Forbes *Crowner's Quest,* pp. 1-6; ERO, D/DP E188 f39.

16   ERO, CR/S1

17   Umfreeville, *Lex Coronatoria,* pp. 102-104.

18   Legal cases are listed in the following order: litigants, date, volume, standard abbreviation of report series, beginning page. *Priestly vs. Fowler* (1837) 3M &W1.

19   E. Jenks, *A Short History,* p. 326.

20 Conformity of the doctrine of common employment with the logic of contract law was made explicit in an equivalent case from 1842 in the United States. See L. Friedman and J. Ladinsky, 'Social Change and the Law of Industrial Accidents', *Columbia Law Review*, vol. 67 (1967), pp. 50-82, esp. pp. 54-6.

21 C. Addison, *Treatise on the Law of Contract and Rights and Liabilities ex Contractu* (London, 1847), p. 745.

22 Ibid., p. 744.

23 'Bailor' and 'Bailee' refer to the maker and receiver of a bailment; a delivery in trust, based on contract, whether explicit or implied, and for a specific purpose. *Fowler vs. Locke* (1874).

24 Historical evidence does not support the liberal view. J. Rule, *The Experience of Labour in l8th-Century Industry* (London, 1981), p. 87.

25 A. Smith, *An Inquiry into the Nature and Causes of the Wealth of Nations* (1776) (Harmondsworth, 1974), Bk. 1, Ch. 10, Pt. 1.

26 H. Broom, *The Philosophy of Law* (London, 1876), p. 79.

27 P.P. 1877, vol. 10, pp. 55 ff; see question no. 1915.

28 Though in practice, the contract-form restricted the scope of these statutory changes. A scan of compensation cases after 1880 shows this limitation clearly.

29 G. Williams and B. Hepple, *The Foundation of the Law of Tort* (London, 1976).

30 *Hadley vs. Baxendale* (1854) 9 Exch. 341.

31 H. Broom, *The Philosophy*, pp. 48-9. Broom refers to *Hadley vs. Baxendale* and to *Hobbs vs. London and South Western Railway Co.* (1875).

32 *Volenti Non Fit Injuria* : 'That to which a person assents is not esteemed in law an injury.' H. Broom, *A Selection of Legal Maxims ...*, 3rd edn (London, 1858), pp. 245-54. On *Volenti* after the Act, see C. Addison *Treatise* , 2nd edn (London, 1883). I will discuss *Volenti* in more detail later.

33 C. Addison, *Treatise* , 9th edn, edited by H. Smith (London, 1892), p. 848; same formulation 11th edn edited by W. Gordon and J. Ritchie (London, 1911), p. 922. On Act and *volenti* defence by employer see p. 925.

34 *Broderick vs. London County Council* (1909) 1 BWCC 219.

35 *Smith vs. Cord Taton Colliery Co.* (1909) 2 BWCC 121.

36 *Robertson vs. Amazon Tug and Lighterage Co.* (1881) 7 QB 598.

37 *Moore, vs. Tredegar Iron etc. Co.* (1938) 31 BWCC 359. These cases are illustrative, and are taken from different periods. They show

a trend in thinking, rather than a historical progression. Many cases are summarized in W. Willis and R. Everett, *Willis's Workmen's Compensation Acts*, 36th edn (London, 1944).

38   *Clover, Clayton and Co. vs. Hughes* (1910) AC 242.

39   *Innes* (or Grant) vs. Kynoch (1919) AC 765.

40   *Turvey vs. Brintons* (1904) 1KB 328; *Brintons vs. Turvey* (1905)AC 230.

41   *Miller vs. Carntayne Steel, etc. Co.* (1935) SC 20.

42   *Cole vs. London and North Eastern Railway* (1928) 21 BWCC 87.

43   *Selvage vs. Burrell* (1921) 1KB 355; *Burrell vs. Selvage* (1921) 14 BWCC 158.

44   *Smith vs. Baker* (1891) AC 325, see p. 360.

45   *Ystradowen Colliery Co. Ltd vs. Griffiths* (1910) 2 BWCC 357.

46   *Dunham vs. Clare* (1902) 2 KB 292; see p. 296.

47   This area is discussed in *Sharp vs. Powell* (1872) 7 CP 253 where the quotation from Addison on torts also appears.

48   Ibid.

49   One of several cases discussed in a review of the problems of intervening agents in a chain of liability; *Clark vs. Chambers* (1878) 3 QB 327.

50   On ideas of causality in legal thought, H. Hart and A. Honore, *Causation in the Law* (Oxford, 1959).

51   Roger Smith has studied the polarisation between will and fact of nature in nineteenth century medico-legal thought, in the context of psychiatric evidence in criminal trials. Extraordinary criminal acts happened by a loss of will and a submersion in nature, in the form of automation. *Trial by Medicine: Insanity and Responsibility in Victorian Trials* (Edinburgh, 1981). I have stressed ordinary, rather than exceptional, events; the acceptance of accidents in a natural realm.

52   S. Huzzard, 'The Role of the Certifying Factory Surgeon in the State Regulation of Child Labour and Industrial Health, 1833-1973', M.A. Dissertation, University of Manchester, 1976.

53   Report of the Departmental Committee on Compensation for Industrial Diseases, 2nd Report, P.P. (1908) XXXV. p. 1, evidence and appendices p. 7 ff.

54   Departmental Committee, P.P. (1913) XVIII, p. 649; evidences and appendices, p. 659 ff.

55   J. Arlidge, *The Hygiene, Diseases and Mortality of Occupations* (London, 1892), pp.2-3.

56   Ibid., pp. 21 ff.

57 D. Blane, 'Inequality and Social Class', in D. Patrick and P. Scambler (eds.), *Sociology as Applied to Medicine* (London, 1982), pp. 113-24; J. Fox and A. Adelstein, 'Occupational Mortality: Work or Way of Life', *Journal of Epidemiology and Community Health*, vol. 32 (1978), pp. 73-8; G. Rose and M. Marmot, 'Social Class and Coronary Heart Disease', *British Heart Journal*, vol. 45 (1981), pp.13-19; P. Townsend and N. Davidson, *Inequalities in Health: The Black Report* (Harmondsworth, 1982).

58 For a historical study of a more social understanding of illness, K. Figlio, 'How Does Illness Mediate ...?'. For a contemporary social theory of illness, examining biographical accounts of health and illness, P. Freund, *The Civilized Body: Social Domination, Control and Health* (Philadelphia 1982).

Part Four: Preventive Policies

WORKERS' INSURANCE VERSUS PROTECTION
OF THE WORKERS: STATE SOCIAL POLICY
IN IMPERIAL GERMANY

Lothar Machtan

This chapter covers the period between the passing of the factory
legislation for the Reich *(Reichsgewerbeordnung)* in 1869 and the coming
into force of accident insurance legislation in 1885.[1] I will consider an
empirical and descriptive report by the Prussian-German state during
the phase of high industrialisation. The state aimed to reduce the
incidence of disease/disability, which became a social problem of great
importance for the first time, and to see that measures to protect the
health of factory workers were satisfactory. These developments are of
interest because the current system of public health services for workers
have in many respects been founded on a crucial historical decision:
the Accident Insurance Law of 1884. At the same time, a system of
preventive measures, based on the examples of other states like
England and Switzerland, was systematically undermined. It is
therefore of value to reconstruct decisive factors in this process: i.e. to
examine which social political discussions preceded these decisions;
how the debate started; which controversies were originally to the fore;
which aspects politics marginalized, or even eliminated before a
specific conception could be translated into concrete laws and then
into socio-political practice.

I The Innovative Phase of State Social Policy

The factory legislation of 1869 can be looked upon as a paradigm of
liberal capitalist economic legislation. It was the starting point for the
state's social measures to protect industrial workers in the *Kaiserreich*.
Its regulations entrusted the form of industrial relations for adult
factory workers to a so-called 'free agreement' between the factory
owner and his workers. There were absolutely no equal rights between
the contracting parties.

In the harsh reality of industrialisation the legislative idea of
employment as an object of 'free agreement' proved to be a *carte blanche*
for entrepreneurial claims to power, which faced the worker as an *octroi*

of obligatory agreements. In this respect the regulations of the Factory Act of 1869 *(Reichsgewerbeordnung)* gave the owners a legal right which entitled them to create positive law within the sphere of their private property. The state sanctioned a sphere within which the entrepreneur could give the constitution of his factory the organisational form of an autocracy. The employer was freed from public and state interference. The possibilities for state intervention were accordingly reduced to a minimum, and were formulated in such a way that effective control could not be expected. These limitations applied to the sole clause for protection of workers which the Factory Act of 1869 contained. §107 stated: 'Every factory entrepreneur is obliged at his cost to produce and maintain all the equipment which is necessary for the protection of the workers against any danger to their lives and health, for the specific kind of factory and its premises'.[2] This clause was not only a provision which could be interpreted in the broadest possible sense, but also a rule which the authorities hardly ever dared observe. The local police formally functioned as a controlling authority, though their officers were neither competent nor prepared to prescribe anything concerning the regulation of public health, since they had not been instructed to by their superiors. Observance of the legal provision was left to the entrepreneurs; the legally required protection of health depended on each factory's policy as determined by capitalist economic requirements.

Government circles became aware the §107 ought not to remain a dead letter when a wave of unprecedentedly extensive strikes began to affect German economic life. The resistance, that a movement of close to 100,000 workers offered, was due to capitalism imposing inhuman pressures at work. Without social protest by the workers, the knowledge of working conditions would hardly have forced a way out of the carefully guarded sphere of the factory and reached the public sphere, where unhealthy working conditions could no longer be ignored. The Social Democratic Party and trade union movement became more influential in the course of the powerful strike movement, thus obliging the government to act.

The beginning of the social political considerations in the Prussian ministerial bureaucracies in 1872 was marked by the realisation that the government lacked reliable information on industrial employment. An elementary precondition for legislative initiatives was not there. To remedy this deficiency the senior officials in the Prussian Ministry of Trade supported the endeavours of the *Verein für Sozialpolitik* (Association for Social Policy) promoting reforms.

The reforming politicians in the Prussian Ministry of State wanted reorganisation of the factory inspectorate. This form of state industrial supervision had existed in Prussia on a voluntary basis since 1854. It only supervised child labour in factories. Subordinate officials were present in only three administrative districts. The inspectorate had become so ineffective that — according to a revision in two administrative districts in 1872 — even the safety provisions for youth only existed on paper. Extensive reorganisation of the factory inspectorate was planned at the beginning of 1873. An organ was to be created:

> by which the government first of all could obtain information on the real conditions of the working classes in factories; secondly, by which it supervises the observance of the provisions for the protection of the life, health and welfare of the workers; thirdly by which the government will be informed of everything which the state and the society can and should do to improve the part of the population doing manual work in a physical, economic, spiritual and moral sense.[3]

Factory inspectors were to:

> supervise the regulations issued for the protection and well-being of the factory workers, and to ensure the effectiveness of the regulations. They are to see to a healthy environment and safe equipment of the workshops, with regard to the lay-out and architecture, and to the working conditions; they are to use their expertise to maintain protection from injurious influences of cold, heat, sweat, dust, poisonous material etc., — to the shielding of dangerous parts of the machines — gears, levers, shafts, belts etc.[4]

These officials were to be responsible both to the government and to public opinion. The qualifications, which their duties entailed, demanded above all personal talents as well as a senior position within the administration. When these criteria were attained in the ensuing period, the question arose whether the supervision of the factory work of children should be removed from the duties of the factory inspectors and integrated into the field of local police administration, so as to leave the factory inspectors more time for industrial hygiene and social statistics. The idea arose of encharging *doctors* with factory inspection, because 'they could take appropriate measures to remedy existing wants because of their inherent experience'.[5] The efforts which the Prussian Ministry of Trade made in a different field of industrial health policy were neither less far-reaching nor less ambitious. In April 1872, after the Ministry had urged the district administrations to finally enforce the regulations of §107 of the 1869 Factory Act through the decree of relevant police orders and not to tolerate the notorious inactivity of the police authorities any longer, it proposed a Bill for industrial accident and disease statistics. With the help of standard questionnaires all accidents causing inability to work for at least 8 days and 'all diseases caused or intensified by conditions in the different factories'[6] were to be reported in every district.

Above all, the officials wanted to know in which branches of industry the workers were exposed to the greatest number of dangers, which kinds of accidents were predominant in specific industries and on which conditions a reorganised factory inspectorate ought to concentrate. In 1873 officials of the Ministry of Trade, accepting proposals by the *Verein für Sozialpolitik* and by the Reichstag, ordered a *Reichstag-Enquete* on women's and children's labour. They strongly recommended such an inquiry not only to the *Reichskanzleramt*, but also exerted a fundamental influence on the formulating of the questionnaire, which for the first time included a section on health.

The verdict on the effectiveness of this package of social reform initiatives is as follows:
1. Most administrations thought the decree of police orders for the protection of the workers to be useless. The orders were a dead letter, as the police authorities forgot about them, and, as entrepreneurs regarded them as tiresome, they tried to avoid them wherever possible.
2. As to the planned inquiry into occupational diseases the idea had to be dropped, because health insurance schemes, factory doctors and entrepreneurs had boycotted the inquiry. 3. The accident statistics did not make progress, because of opposition from the entrepreneurs. Minor accidents happened almost daily — as the Chamber of Commerce of Essen noted in 1875 —, and so the factory owners felt that it could not be expected of them to report the details of all these events. 4. Ultimately, the answers to the questions of the *Reichstag-Enquete* on women's and children's work vividly illustrated how injurious to health most industries were which employed women. The *Enquete* also documented an unmistakable scepticism about the effectiveness of a reorganised factory inspectorate, although the inability of the local police authorities concerning the protection of the

workers was not denied. It was the administration of the most industrialised district of the Prussian monarchy — Düsseldorf — that gave the clearest warnings; they voted for industry, which should be able to develop according to its own rationale, and recognised that interventions into its inner organisation would only be tolerated unwillingly if at all.[7]

The conclusions drawn by the Ministry of Trade from the reports were that it was necessary to complete the legal amendments to the 1869 Factory Ordinance, and to reorganise factory inspection, but initially only in Prussia. The way that the latter proposal was put into practice justifies the view that the bureaucratic energy to reform had been already dampened by 1874/75, because the high-ranking administrators accepted the reservations of industry against a too rapid extension of industrial supervision and against the employment of doctors as factory inspectors. From 1874 to 1877 eleven new factory inspectors' posts were created for the whole of Prussia, and they were exclusively for chemists and technicians. As to the powers of the officials, their instructions showed clear signs of concessions to the factory owners. Their representatives had suggested that the government organise industrial supervision in such a way that the entrepreneur regarded the factory inspector 'as an advising friend... but not as an opponent'.[8] Although the new supervising official was for the first time entrusted with state assistance to enforce §107 of the Factory Act by inspection and instruction of the police authorities, the possibilities to intervene were so restricted that in practice he only had an advisory and observing role.

The reforming group's socio-political ambitions within the Prussian government were not discouraged by this opposition, and they tried to realise their plans through a legal amendment to the factory legislation. At the end of June 1876 the Prussian Minister of Trade already presented a bill to the *Reichskanzleramt* and to the Ministry of State which was based on the reform proposals of 1872/73. It contained not only a number of stronger regulations to protect women and children factory workers, but also included an amended version of §107, as well as the obligatory introduction of factory inspectors on the level of the Reich. (These clauses are given as Appendix 1 of this paper).

Without exaggerating one can say that this Bill was a flight of fancy of state social thinking within the Prussian state's bureaucracy, and a testimony to socio-political principles, that stressed preventive measures. Above all this Bill was the first state political negation of the liberal axiom of a factory as a non-public, and more or less secret realm beyond the reach of the law. The Bill only provided *possibilities* for intervention, but these were nevertheless suited for ascertaining information and events, which had been previously hidden from public power. It was a starting point for effective industrial protection of health. In fact, the Bill was anything but radical with regard to the principles of capitalist production — as the appendix belonging to it clearly showed; its authors declared that they were concerned only with the 'realisation of moral principles' in factories as a hitherto isolated part of the social community. The legal regulations were intended to open the eyes of the entrepreneurs to their duties and, by means of industrial supervision, to convey the idea of the state as opposed to one-sided class interests. These aims were linked to halting the growing influence of Social Democracy.[9]

It should finally be observed that the Bill could be compared to the factory legislation in other industrial states like Switzerland or Great Britain. It could certainly rely on broad Parliamentary agreement, particularly since it took into account the demands of the social reformers of the Conservatives, the Progressive Party, the (Catholic) Centre Party and even part of the National Liberals.

II The Restrictive Phase

The indirect blow struck at the as yet almost unlimited freedom of the entrepreneurs provoked a protest within the ranks of this class. They succeeded in deciding the fate of the Bill within a few months, and thereby in initiating a new course of great importance.

The starting point was a critique of the Bill by Alsatian industrialists, at the request of the *Reichskanzleramt*. The industrialists categorically declared that the planned protective regulations exceeded by far anything that the competitiveness of the German industry could bear. Besides the impracticable regulations, the Bill contained a number of highly unnecessary burdens to the employers, as for example, the duty to report an accident. If the proposed regulations were enacted, they would damage the interests of industry so that it would be unable to compete and would thus be ruined in the long term. Eight days after this report was sent to all Prussian Ministers of State, Bismarck took up the criticisms voiced by the industrialists and wrote:

> In view of the current adverse situation of industry,
> every legislative action which aggravates production
> without important reasons seems to be critical.
> Industry needs a rest to be able to take up the
> struggle against foreign competition again after the
> recession of the last years. The current point of time is
> unsuitable for modifications to our legislation, which
> will disturb the running of the factories or will be
> disadvantageous to the efficiency of industry at home
> rather than abroad.[10]

This new position of the government on workers' protection was announced almost *ex cathedra* to the Reichstag in the spring of 1877, when the *Reichskanzleramt* President, Hofmann, declared that the government did not intend 'to introduce reforms of the trade regulations, which would interfere with the running of a factory, or to implement decisive new norms'.[11] The government was in complete agreement with the economic pressure groups and their parliamentary representation. The state now carried on a propaganda campaign so as not to lose the confidence of industry because of socio-political experiments.

In the course of 1877, Bismarck, after having consulted influential industrialists, strengthened his opposition to social reform plans of the Ministry of Trade, especially on the issue of public health. He opposed the idea of transferring to the factory inspector the monopoly of definition as to what was necessary to secure the life and health of the workers as in §107. By doing this the government would create unnecessary enemies among the 'influential class of factory owners', because the latter felt threatened in their freedom of action. He had no understanding for 'why among all branches of human activity it was just with the most difficult and most vulnerable to foreign competition that the protection against dangers threatening human life should be extended to such a degree as in §107'. Institutions that promote health in factories would require such extensive and costly facilities 'that only rarely and if unusual profit prevailed, entrepreneurs would want them'. He concluded that 'every new obstacle and artificial restriction in factories reduces the employer's ability to pay wages', and this the state could not tolerate.[12]

This ostentatious withdrawal from the socio-political conception of prevention in favour of a pretence of securing of jobs was classically formulated in the Reichstag speech in 1883 by Theodor Lohmann. He had advocated exactly the contrary ten years before and, despite his better knowledge as an authorised agent of the government, had to declare: 'However important the aim of protecting the workers against dangers, the necessity — whatever it may be — remains relative. The aim is by no means one that unconditionally and under any circumstances had to be and could be reached...it is still more correct to maintain the worker's basis to exist, while he cannot be relieved of certain dangers, than to completely withdraw the basis of his existence'.[13] The logic of this social principle was that in the years after 1877 the government tried to reduce rigorously all starting points for protection of health for the workers with the help of the technical and managerial expertise of the associations of the entrepreneurs . Thus it tried to remove the essential barriers to what had become only a token solution to the problem.

The first step in this direction was the changing of the factory inspectorate into a state advisory body, responsible for improving the safety of machinery. It reversed the decision by the Reichstag in 1878 to turn the inspectorate into an obligatory institution, as it was against the wish of the government. As an advisory body, the inspectorate was forbidden to 'hinder or damage industry in its use of machinery and of other methods, and generally in its free development'.[14] This reduction of the protection of health to merely technical protection against accidents can be compared to leaving industrial hygiene out of the tasks of industrial supervision.

Entrepreneurial organisations similarly prevented the decree of detailed practical regulations concerning the protection of public health in the Reich trade code. The Reichstag had also handed over competence in this sphere to the upper-house (*Bundesrat*). The influential industrial associations found it easy to convince the government with arguments against preventive measures in this field. They maintained that it was impossible with the current level of technology and with its rapid progress to make regulations, which were applicable to diverse branches of industries. The consequences of the decree would be manifold disturbances of the factories, disputes with the supervising authorities (which could not be expected to have a full understanding of conditions in different branches of industry) and with the workers, avoidance of necessary precautions and frequent denunciations on the part of the workers, and, finally, in case of an accident unjustifiably heavy responsibility of the head of the factory who had not kept to the letter of impracticable regulations. To make every accident in a factory absolutely impossible, would mean making the factory itself impossible. Factory legislation ought only to protect workers as much as was practicable.

The socio-political policies of the Bismarck government became fixed in its principles of 1877. In his programme of 1880 for compulsory social insurance, the law for accident insurance (passed in 1884) was central. The various bills which the *Reichskanzleramt* drafted on this theme in the years from 1881 to 1884 were by no means due to the political genius of the *Reichskanzler* but much more a compilation of suggestions which had been under discussion since 1878. This shows how much the government had, when making decisions, to rely on the help of those interest groups, to whom it had felt particularly obliged with its economic and social policy since the end of the 1870s. This can be seen in the central features of the Accident Law. Accident insurance, the principle of mutuality, and the trade co-operative organisational form had been frequently discussed among the ranks of the entrepreneurs, the private companies of accident insurances, and the supervising officials. They had been presented to the Chancellor in an oral and written form, before he had them transformed into legal terms. This also applied to the intention of transferring part of industrial supervision, i.e. the regulations regarding the industrial protection of health, to trade co-operative associations and thus transferring them into private ownership. In the preamble to the Bill of 1884, these motives were openly laid down. (See Appendix 2)

Bismarck thought of the Bill as having a socially integrative function in the sense of the state supporting tendencies, which were to secure the participation of independent workers' commissions in the examination and prevention of accidents. But the Bill did not come into force because of lobbying by entrepreneurs. The participation of the workers became a mere decorative appendix to the policy of entrepreneurial interests. The industrialists alleged that the danger of the workers' commissions was that 'the subordinate would be organised against his superior' and would become aware 'that his interests collided with those of his master'.[15] The industrialists furthered the principle of a social and health policy, which resulted from economic calculation. The elimination of the workers' commissions from the Accident Law meant renouncing the valuable knowledge which the workers could have added to accident prevention because of their practical experience. It also meant leaving out moral and hygienic issues as well as those relating to standards of living from any concept of health. Compared to the material interest of the entrepreneurs for effective prevention of accidents, the action of the *Reich* insurance authority under Bödiker, aiming at extensive measures of preventing accidents, was hardly anything other than a token corrective. A decrease in numbers of accidents in connection with this action cannot be proven.

Insurance compensation against risk and the simultaneous transferring into private ownership of the prevention of risk meant the transference of the combating of causes to entrepreneurial private initiatives. This direction of state social policy, which started with the insurance against accidents, was the first basic step dividing the combatting of causes (diseases of workers, occupational diseases) from the tasks and terms of reference of state social policy.

If one searches for historical alternatives to the process outlined on the formation of the health-specific part of state social policy, the answer is that there have been no models which pointed beyond the state model in 1876. The medical profession did not provide organised facilities for workers' health until the end of the 1880s. The doctors, who dealt with the questions of industrial hygiene and occupational diseases, understood their practice mainly in a positivist scientific sense, i.e. as an enrichment of medical knowledge. They mostly renounced any generalisation of their isolated observations concerning public health. But even the few doctors who did not shrink from making explicit socio-political remarks did not go beyond the demand of moderate additions to the factory legislation as well as successive extension of industrial supervision. They did not think of any extensive primary prevention. At the fifth annual meeting of the German Association for Public Health in 1877 the main address by Beyer, when warning of the exaggerated demands concerning public health, emphasized his conviction, 'that precisely in this field prudence and precaution are necessary, and that one should not try to support a hygiene which risks or neglects the main priority, the securing of the daily bread. That is why it is necessary to openly oppose those immoderate demands which hide behind their pleasant mask of hygienic and humanitarian ends quite different aims. One should not let emotions reign but considerations and experience'. Or, in the words of the district medical officer, Merkel, 'hygiene must not emerge as the enemy of industry'. [16] It is striking that the demand to employ particular industrial doctors was put forward by social reformers and not by doctors when it was made for the first time in 1891.

It is appropriate to conclude with some remarks on the ideas of the protection of workers of the early Social Democrats. They advanced beyond the state reform model of 1876 only with regards to the principle of protection, and not regarding the extent and the contents of the protective regulations themselves. (The questions of the regulation of working hours or the average working day are not taken into account here). Their main concern was the question of the organisational form of the industrial supervision . In their plan of 1877 they demanded that the *Reichsgesundheitsamt* (Board of Public Health), established shortly before in 1876, should nominate the factory inspectors because, 'precisely the professional class represented in the *Reichsgesundheitsamt* has on the average retained its political independence and can be characterised by humanity and true, working class philanthropy'.[17] As an organisational model the 1885 Bill for the Protection of Workers intended the establishment of chambers of industry with representation on the principle of parity on the level of the factory, as well as the establishment of a labour exchange as a control authority. In other words, the Social Democratic reform proposals basically aimed at a democratisation of the industrial *administration* in the form of participation of the workers, but they did not aim at a fundamental restructuring of state social policy in the form of primary prevention of health risks.

Appendix 1

Regulations from the Workers' Protection Bill of 1876 drafted by the Prussian Ministry of Trade.

107. The rooms, machinery and other equipment in every workshop are to be so installed and maintained that the lives and health of workers are effectively protected against hazards.

Especial care should be taken in the provision of adequate ventilation, light and for the expelling of dust, vapours and gases arising during manufacturing. There must also be measures to protect the worker against dangerous contact with machines and machine parts and against other hazards arising in any occupation or factory.

107a. In every workshop good morals and observance of public decency must be maintained.

...

107c. The police are to see that the facilities as defined as necessary in 107 and 107b are provided in individual cases, and that these are implemented according to the capacity of a place of work.

107d. The owners and managers of all places of work as defined in 107 and 107b are required to notify to the police all accidents that occur in their places of work to workers in their employment, whenever the injured person or persons do not return to their place of work within 24 hours.

The report must contain in addition to an exact description of the workshop, the time and type of accident, number of injured and the nature of the injuries.

...

132. The supervision over the facilities as defined in 16 and to enforce the regulations of 107-107e and 128-131 (factory inspection) is exclusively vested in the regular police authorities and in special auxiliary officials.

132a. The officials of the factory inspectorate are to remain under the authority of the regular police in their allocated districts with regard to all their powers, but excluding the right to decree police orders. (the code of laws determine the powers they have at their disposal, and the procedures for prosecutions and appeals).

132b. The officials of the factory inspectorate must, after showing their warrant, be granted access to all parts of a workshop instantly and at all times.

132c. The officials of the factory inspectorate are empowered to question the owner and managers and all persons employed about all events and conditions that occur within their spheres of responsibility.

132d. The local police are required to provide the factory inspectors on their visits all documents about the workshops in their area, and to give assistance as defined in 132, and to make available all documents on any enquiries or legal procedures taken.

The owner and manager of the facilities in question are responsible for the observance of this regulation by any persons in their service who are responsible for supervising access.

Appendix 2

The Reich Accident Insurance Law of 1884: Extract from the Preamble.

The formation of *Berufsgenossenschaften* [as a special kind of employers' association to finance accident insurance] is the only means of attaining an effective system of accident prevention because the transfer of accident insurance to such corporate bodies corresponds to the historical development of associations in the social and economic spheres. It can hardly be doubted that police intervention on its own cannot satisfactorily fulfil these immense responsibilities, and consequently participation of those involved is indispensable. The greater or lesser degree of danger in a factory should determine the amount due for compensation to be provided by the *Berufsgenossenschaften*, as has been suggested by the *Reichstag* Commission for the discussion of the previous draft of this law. There results a substantial financial interest to prevent accidents, both on the part of the employer and of the employers' association. For economic reasons and in the interest of the general public, it is necessary to allow autonomous administration greater freedom of play. On the one hand the employers' associations must attempt to improve the facilities in places of manufacture, and to prevent effectively accidents by using as far as possible the latest technology and other appropriate means. On the other hand it must be recognised that only completely proven equipment can be prescribed as necessary, and that costly experimental equipment should be avoided as placing too heavy a burden on industry. Industry must be protected against mistaken interventions, however well meant, when these would jeopardise the profitability of a particular branch of industry or disregard the competiveness of entrepreneurs. The employers' organisations will succeed all the sooner in arriving at a judicious standard, when they take into account the viability of the individual entrepreneur and the position of a whole branch of industry, so that both interests can be balanced.

Notes

1   For references to the general literature see the notes to the chapter by Dietrich Milles.

2   *Fabrikgesetzgebung des deutschen Reiches und der Einzelstaaten* (Berlin, 1873), pp. 43-44.

3   Memorandum of the Geheimer Regierungsrat Jacobi, quoted in the journal *Concordia*, 13th March 1873.

4   Ibid.

5   Deutsche Reichskorrespondenz, quoted by the socialist

newspaper *Neuer Sozialdemokrat,* 12th February 1875.

6   Decree of the Prussian Chamber of Commerce, 30th June 1873, Hauptstaatsarchiv Düsseldorf, Reg. Düsseldorf Nr. 24583.

7   Letter from the provincial government in Düsseldorf to the Prussian Chamber of Commerce, 16th February 1874, loc.cit.

8   Remark by the factory owner Kalle during a speech in the Prussian *Abgeordnetenhaus* 29th February 1876 (*Protokolle,* p. 345).

9   Hans Rothfels, *Theodor Lohmann und die Kampfjahre der staatlichen Sozialpolitik* (Berlin,1927), p. 32.

10   Directive of Bismarck, 30th September 1876 cited by H.v. Poschinger (ed.), *Aktenstücke zur Wirtschaftspolitik des Fürsten Bismarck,* vol.1 (Berlin,1890), pp. 233-234.

11   *Stenographische Berichte über die Verhandlungen des Deutschen Reichstags,* (12th March 1877), pp. 95-96.

12   Letter of Bismarck to the Prussian Chamber of Commerce, 10th August 1877, Zentrales Staatsarchiv Merseburg Rep. 120 BB VII 4 Nr.1 Bd.2.

13   *Stenographische Berichte...* (12th January 1883), pp. 864-65.

14   *Jahres-Berichte der Fabriken-Inspectoren für Berlin und die Provinz Schlesien für das Jahr 1874* (Berlin, 1875), p. 27.

15   Opinions of businessmen regarding parliamentary proceedings on the first *Unfallversicherungsvorlage* 1881, quoted by Monika Breger, *Die Haltung der Unternehmer zur staatlichen Sozialpolitik 1878-1890* (Frankfurt/M., 1982), pp. 119-124.

16   *Deutsche Vierteljahresschrift für Öffentliche Gesundheitspflege,* vol.10 (1878), p. 154.

17   See August Bebel, 'Das Reichs-Gesundheitsamt und sein Programm, vom socialistischen Standpunkt beleuchtet', *Die Zukunft. Socialistische Revue,* vol.1 (1878), pp. 369-70.

18   A. Lohren, *Entwurf eines Fabrik- und Werkstatten-Gesetzes zum Schutz der Frauen- und Kinderarbeit, hergleitet vom Standpunkt der ausländischen Konkurrenz* (Potsdam, 1878), pp. 78 passim.

19   *Stenographische Berichte...* (1884), enclosure no.4.

12 AN INSPECTOR CALLS: HEALTH AND SAFETY AT WORK IN INTER-WAR BRITAIN

Helen Jones

In Britain the subject of occupational safety and health has received little scholarly attention; interest has centred on the nineteenth century and on the two world wars when a close correlation was widely recognised between maximum production, good industrial relations and good working conditions. Here, our attention will focus on the years between the wars; it was at this time that much of the research was undertaken which formed the basis of later action by both government and industry: the Industrial Health Research Board and Institute of Industrial Psychology both produced a steady flow of findings. Their activities were related to changes in the labour process brought about by the permeation of Taylorist ideas into Britain, and it was a period when the number of employers interested in safety, health and welfare as an aspect of labour management grew on an unprecedented scale. The voluntary nature of this activity meant that inequalities in standards of safety and conditions of health between firms and regions increased as the legal standards remained those laid down in the 1901 Factory and Workshop Act.

There was one group working within this environment which was consistently interested in safety and health at work and whose starting point was not an industrial relations but a safety, and to a lesser extent, health one. Since its inception in 1833, the factory inspectorate had been the linchpin between state and industry in occupational safety and health; during the inter-war years new channels of influence, such as tripartite committees in industry and the Industrial Museum, opened up, and the inspectorate solved problems which during the two world wars were taken charge of by politicians.

Most of the research and debate concerning the factory inspectorate has concentrated on the nineteenth century. This is of limited value to the historian of twentieth-century occupational health: the dominant values of society differed, and to focus on the work of individual inspectors, as many of the writers on the nineteenth century have done, is inappropriate, for by the twentieth century certain of the inspectors' extensive discretionary powers had been curtailed and a greater uniformity imposed, while the nature of the extant records for the twentieth century makes such an approach well nigh impossible.

This chapter discusses the role of the factory inspectorate in improving safety, health and welfare standards and in shaping the climate of opinion over what were 'acceptable' basic standards. Such a comment immediately raises the problem, already noted by Eugene Bardach, that much discussion of implementation must inevitably be impressionistic; implementation is not concentrated in one place, but is dispersed, so one either provides a narrow view with data or an overview where rigorous and systematic methods are not feasible.[1] Here we adopt the latter course. Our argument will be developed by discussing the inspectorate's methods of inspection; its work and influence outside the scope of the law and its influence on the law, and finally the impact of the inspectorate on workers' attitudes towards safe and healthy methods of work.[2] We conclude that current concepts of factory inspection are inadequate and an alternative one is suggested.

I

The method and efficiency of factory inspectors is a source of debate among historians and sociologists. P.T. Fenn and P.W.J. Bartrip explain what they perceive to be a lack of efficiency in nineteenth century inspectors in terms of fiscal restraints imposed by the Victorian community.[3] W.G. Carson argues for a broader and more rigorously defined conceptual framework; in his work on factory inspectors in the 1960s he argues that their methods of enforcement were a means of achieving functional efficiency.[4] Previous discussions of the factory inspectorate have emphasised the educative role of the inspectorate, but have nevertheless retained an underlying commitment to placing the inspectorate within a 'law enforcement' framework. This chapter will argue that a more appropriate definition of the function of the factory inspector is a factory 'instructor' or 'teacher'. Such a definition encompasses a law enforcement role while embracing much else. First, inspectors transmitted values and standards; they emphasised the importance of a safe and healthy working environment and suggested what this meant in practical terms. Second, inspectors provided a resource of knowledge and skills which was the basis for their acceptance by employers and workers (it was from this, rather than from statutory powers as Gerald Rhodes suggests, that their real authority derived). Third, inspectors evaluated the safety and health of a factory, detected bad practices and judged the appropriate action to be taken. By these means they were able to act as instructors and socialisation agents; much of the inspectors' work aimed at self-regulation for industry.

Moreover, the legal aspects of the inspectorate's work have been somewhat misunderstood. The legal enforcement of the Factory Acts falls into the general pattern of law enforcement even though the enforcement agency was the Home Office factory inspectorate, not the police. This has not always been fully recognised. First, historians and sociologists, by emphasising the extensive discretionary powers of factory inspectors, imply that this is peculiar to, or at least more prevalent among, factory inspectors. Bartrip and Fenn, working on factory inspectors from their early years to the 1860s, comment on the great variety in the number of prosecutions from one district to another: inspectors had the power and ability to vary policy in their districts.[5] Carson, analysing the factory inspectorate a century later, in the 1960s, argued that the decision to prosecute depended on the individual inspector's interpretation of the offences and his or her personal assessment of the firm's overall attitudes and policy.[6] Yet, the extensive use of discretionary powers is also typical of other areas of law enforcement;[7] the need for discretion and selective enforcement is recognised by the police.[8]

Second, there is an emphasis on the infrequent rate of prosecution among employers for 'white collar' crimes. Carson and Wiles write that while the known amount of violation of the law by owners and occupiers is substantial, prosecution is infrequent and whether by accident or design this group is thereby grossly underrepresented among officially defined offenders.[9] Carson, writing specifically about violation of Factory Acts, stated that it was countered almost exclusively by the use of formal administrative procedures, other than prosecution.[10] Yet, here again, the unwillingness to prosecute employers, except as a last resort, reflects enforcement agencies' lack of confidence in the Courts and the inappropriateness of the use of the Courts in many cases.[11] There is a conflict between what the Courts and what the enforcement agencies regard as 'justice'. This, rather than Carson's concept of 'institutionalised tolerance' helps to explain the low rate of prosecutions.[12] While the factory inspectors tolerated certain breaches of the Factory Acts this tolerance was highly selective and it was not a reflection of the homogeneity of legislators, judges and administrators with businessmen, as Edwin Sutherland argued in his seminal study of white collar crime.[13] The prosecution rate of all researched non-police law enforcement agencies is low.[14]

Placing factory crime in the mainstream of law enforcement raises its own difficulties of analysis as present emphases in criminology are unhelpful. Quinney, for instance, argues that, 'criminal definitions describe behaviours that conflict with the interests of the segments of society that have the power to shape public policy',[15] which implies that those who are able to influence the law are less likely to commit crimes. Businessmen had the power to influence, although not dictate, the nature of factory legislation, yet still not all businessmen, at all times, complied with the law. It is also difficult to explain employers' lawbreaking by reference to concepts such as inadequate socialisation, criminal subculture and criminal motivation.[16]

While creating problems for the historian, the absence in the 1920s and 1930s of either academic research or popular notions of widespread crime among employers must also have hindered the work of inspectors, for the definition of an activity as 'criminal' has a useful propaganda purpose.[17] Yet, at the same time the widespread flouting of one or more sections of the Factory Acts reduced the stigmatising effect of criminal labelling.[18] These problems were compounded by a variety of other factors which detracted from a full implementation of policy, varying over the period in their intensity: these included size and competence of the factory inspectorate; contemporary views on inspection; diversion of resources and deflection of policy goals; resistance to explicit and usually institutionalised efforts to control behaviour administratively; dissipation of personal and political energies and ambiguities over methods of enforcement and interpretation of the law.[19]

Factory crime was well suited to persuasion rather than prosecution for factory inspectors were much concerned with changing people's attitudes. Their aim was to encourage an awareness of the hazards of work.[20] This underlines the positive nature of much of the inspectorate's educative work: it was not merely an alternative to a time-consuming and often hostile criminal process, but it was actually a more appropriate means of attaining certain ends. During the 1930s the inspectorate aimed at achieving, as far as possible, a system of self-regulation in industry, and education was a means to this end; criminal sanctions were merely a technique.[21]

Employers who broke the law were not indulging in activities which posed a threat to 'law and order'; their behaviour did not lead to social unrest among either themselves or their victims. In part, this was because employees' discontent was institutionalised and either channelled through the trade unions or direct representation was made to the inspectorate; it also reflected the lack of a clear division between employers and workers, for many of the latter also broke the factory law.[22] That many accidents resulted from workers flouting the law, neglecting to use safeguards provided by the employer and disregarding instructions as to the manner in which work should be performed was a frequent refrain of inspectors.[23]

Prosecutions were undertaken as a last resort; indeed one senior inspector commented in 1934 that she came across members of the inspectorate who after five or six years' service had not undertaken a prosecution.[24] Before undertaking prosecutions, inspectors took into account the general attitude of the firm towards safety and health. Firms were well aware of this, and on one occasion, at least, brought pressure to bear on the Home Office not to prosecute a firm with a good safety record.

Other considerations also affected the prosecution rate. While an inspector might secure a conviction against an employer this could rebound to the disadvantage of the worker. There was no means of ensuring that a worker was not victimised following action by an inspector. In fact, employees could be dismissed on the mere suspicion that they had contacted a factory inspector. This was an inducement to inspectors to educate employers in good practice and to avoid prosecutions.[25]

There was little faith in the criminal system's ability to deal with criminal activity under the Factory Acts: inspectors devised means of circumventing it and only prosecuted in a small percentage of instances; the legal profession was disparaging of magistrates' ability to deal with cases under the Factory Acts and the labour movement was disgruntled with an appointments system which appeared politically biased.

When a conviction was secured the fines imposed were often derisory, and not worth the time spent by inspectors on the case.[26] Fines were too low to act as a deterrent and there was no reformative element in the law. While magistrates who had visited the Industrial Museum frequently alluded in Court to the practical help the museum could give in complying with the Acts, such visits were not obligatory.[27]

Moreover, ambiguities in the law made it difficult for inspectors to take up a case with any certainty that they would win it. Indeed, the attitude of the Bench was widely regarded as a major obstacle to law enforcement. The Lord Chief Justice repeatedly castigated the magistracy for their decisions. One case in 1937 stands out. The King's

Bench Divisional Court had allowed a factory inspector's appeal against the dismissal by J.P.s of a case which arose out of an accident to a boy injured while working at a power press. The J.P.s had found that the machine was dangerous and unfenced; it could have been securely fenced and if it had been, the accident would not have occurred, but the machine was as safe as if it had been fenced. The Lord Chief Justice commented, 'This is the kind of case which almost makes one despair', and he admitted that he was frequently tempted to comment on the behaviour of some lay magistrates when they were trying Factory Act cases,

> They give the go by to the plain meaning of the statute...come to contradictory conclusions, and arrive at an opinion against the prosecution. In the last fifteen and a half years I have had so many of these cases in which the same fallacy and the same blunder are cheerfully repeated, that I am coming to the conclusion that sooner or later somebody will have to propose a clause in an Act of Parliament which will provide that all prosecutions under the Factory and Workshop Acts must be dealt with by justices of legal training and independent position. In this case the justices have flatly contradicted themselves and have come to a conclusion which but for the respect we have for all justices I should describe as 'grotesque'.[28]

In Parliament, too, there were questions about the 'independence' of J.P.s in Factory Act cases. In 1932, Rhys Davies a Labour M.P. complained about the low level of fines and suggested that the reason might be that there were too many employers of labour on the magisterial bench dealing with Factory Act cases; he urged that a much more serious view should be taken of infractions of industrial laws.[29] In fact, nearly half of all magistrates were employers, while only about one sixth were 'wage earners'.[30] Nineteenth century historians have referred to the occupational bias of the Bench, but then this reflected the formal political power structure in the country as a whole.[31] The Labour party gained Office briefly during the inter-war years, and the trade unions were involved, to some extent, in the policy-making process for safety and health at work, but for the actual enforcement of the law this was irrelevant. During the 1930s the power of organised employers was curtailed in policy-making for industrial social legislation, yet employers' influence at the level of implementation remained strong.[32] While it was possible to have the decisions of J.P.s reversed, this protracted the criminal process and brought it into ill-repute. The criminal process reflected, in a harsh fashion, the real distribution of economic and political power in inter-war Britain; by doing so, in such a blatant manner, it brought the condemnation of even those who supported the structure.[33]

II

From the earliest days of inspection there had been an emphasis on educative work, but from the late 1920s it was practised through new formal organisations. First, an Industrial Museum, opened in London in 1927, was run by the inspectorate. It was an exhibition of methods, arrangements and appliances for promoting safety, health and welfare using equipment on permanent loan from firms.[34] Local exhibitions were also organised.[35] The impact of the museum was immediate. One small safeguard caused such interest that its makers opened a factory for its manufacture. Technical schools were prompted into fencing their demonstration equipment on lines seen in the Museum and one district inspector received a letter from a firm which in the past had shown a 'marked reluctance' to comply with the Factory Acts,

> I have seen this Industrial Museum and I was very much impressed with what I saw there...It has just made me wonder whether our works are up to date in safety first appliances. We have not much money to spend because [this] trade is going through an awfully bad time, but you might just walk round when you can find a few minutes to spare and see if we are up to date...I now see what the Inspector of Factories can do with a little warm-hearted attention to firms with not so much officialism, and you can destroy some of my previous correspondence.[36]

The Chief Inspector felt that the Museum raised the status of the inspectors in the eyes of managers; one manager told the inspector for Bethnal Green that before his visit he had maintained that inspectors dealt with such a variety of industries that they could not possibly know what they were talking about in connection with any one, but since visiting the Industrial Museum he realised that as Inspectors saw so much and had so much information, they must be exceedingly well qualified.[37]

Following a conference between members of the Association of Technical Institutions and Inspectors held at the Museum, the inspectors began delivering lectures on safety at technical schools. (Often in addition to students, local employers and workers were invited).[38] Lectures were also delivered to individual works. In 1933 the Factory Department supplied the T.U.C. Annual Conference with slides and a lecturer on the Industrial Museum, and trades councils organised Home Office lectures in fifteen centres. To accompany its lecture work the Home Office published a number of safety pamphlets.[39]

Factory inspectors also tried to raise industry's standards through the traditional channels in which British industrial relations are conducted, that of collective bargaining within industry. By their representation on firm- and industry- based committees the inspectorate established a foothold in the formal structure of industry. In 1931 the Chief Inspector expressed the hope that permanent standing committees to review safety would be formed in all the major industries. He envisaged the Factory Department becoming involved in the work of these committees.[40] Already, in fact, inspectors were involved in works' safety committees: they attended meetings, delivered lectures and revitalised waning committees. They were also members of Local Area Safety Committees comprising industries to which a Draft Safety in Factories Order applied.[41] Between 1926 and 1928 employers, workers and the factory inspectors negotiated agreements over the fencing of machinery, the prevention of accidents and first-aid in cotton-weaving factories. A Joint Standing Committee, comprising representatives of the Cotton Spinners' and Manufacturers' Association, the Federation of Master Cotton Spinners' Association Ltd., the Northern Counties Textile Trades Federation and the Factory Department met twice yearly to consider amendments to the agreements under the chairmanship of the Superintending Inspector for East Lancashire.[42]

The efforts of the Factory Department and the employers' organisations to reduce accidents by voluntary action demonstrates the lack of a clearly defined role for the state in occupational health. While the T.U.C. and Labour party demanded state intervention, the employers' organisations took up the question of voluntary action to reduce accidents.[43]

The specialist, medical, engineering and electrical inspectors contributed much to the dissemination of knowledge and the voluntary adoption of safe and healthy practices in industry. The work of the medical branch had been enhanced since the First World War when it was shown quite emphatically that the health of the worker was of great economic value. The Chief Medical Inspector, Sir Thomas Legge, who had an international reputation for his expertise in the field of industrial diseases, made enquiries during the First World War into the industrial chemicals which caused jaundice and aplastic anaemia. Dr. E. Middleton, before joining the Department in 1921, published an article on the occurrence of silicosis. He was the first to carry out a systematic clinical survey in the grinding industry, or in any industry producing a pneumoconiosis, and demonstrated quite clearly the incidence of silicosis among these workers. This research was to form the basis for further enquiries into other industries in which workers were exposed to the dust of silica, in particular in the sandstone and pottery industries. These enquiries gradually gave rise to schemes for compensation and periodic medical examinations carried out by the Silicosis and Asbestos Medical Board.[44] The 1937 Political and Economic Planning report on the health services thought that the medical inspectors did 'invaluable' work, but there were too few of them.[45] According to an assessment made by the Consulting Medical Officer for the General Electric Company, Donald Stewart, in 1941, the medical inspectors, over the previous decade and a half, had been helpful to medical officers entering industry and to employers who introduced medical and nursing services in their factories. Stewart felt that because of their extensive knowledge and experience medical inspectors played an important part in the prevention of occupational diseases and much of the success in eliminating such diseases from industry had been due to their efforts.[46]

III

The great store placed on persuasion and educative work by the inspectorate meant that their credibility and the extent of confidence felt in them by the workforce were of vital importance. The attitude of trade unions towards the inspectorate can be gauged from the reports of trade unions to the T.U.C. in 1929 when it was collating evidence to submit to the Departmental Committee on Factory Inspection. These reports, which only reflect attitudes at one particular time, can be supplemented by T.U.C. Annual Reports.

Among trade unions a consensus existed as to the desirability of the system of factory inspection. Almost every union urged more frequent inspection and pinpointed much of the weakness of the system on the infrequency of inspection. For instance, the Amalgamated Weavers' Association alleged that there was a considerable amount of time-cribbing by cotton employers, which in the union's view was due to a lack of inspection. This inefficiency was seen not as the result of an unwillingness among factory inspectors but as their physical inability to cover the ground in the requisite time. In Blackburn and Burnley it was 'impossible' for the staff to pay adequate attention to inspection, investigation and prosecution.[47]

While favouring more frequent inspection there was a general desire that the inspectorate should be invested with greater powers. The National Federation of Building Trades Operatives wanted the inspectorate's powers increased in two specific ways: where an inspector found any faulty material which was likely to be used surreptitiously he should have the power to have it immediately destroyed, and if he found defective material being used in the erection of scaffolding he should have the power to order its dismantling. If the management failed to comply with inspector's instructions the inspector should have the power to stop work with the defective material or on defective scaffolding.[48]

Despite many glowing reports on the inspectorate and its work and with no suggestion that the inspectorate was the tool of the capitalist system in general, or the employers in particular, when a Senior Inspector of Factories drew up plans for a national safety campaign the T.U.C. objected that it was a scheme to protect employers' property at the workers' expense. The proposed annual subscription of 6d per year in conjunction with a rule 'To endeavour to prevent damage to the property, plant and machinery of my employer' caused particular annoyance.[49] Despite the recognition of a common interest between trade unions and factory inspectors, specific attempts to co-operate closely could founder on the absence of an agreed framework within which to work.

The support trade unions professed for the inspectorate suggests that the official union line was a condemnation of criminal activity in the field of safety and health at work. Certainly, the workforce did use the inspectorate when employers committed crimes. In the late 1920s the United Ladies' Tailors' Trade Union provided the Home Office with examples of insanitary workshops in London. The Factory Department immediately took action and inspectors visited workshops in East and West London. Extensive improvements were ordered in lighting, heating, space, ventilation, lavatory accommodation and general cleanliness. The union announced that a number of 'infested bugholes' had gone out of existence after an inspector's visit.[50] Between 1921 and 1931 the Annual Reports of the Chief Inspector of Factories detailed the number of complaints received in a year about contraventions of the Factory Acts: the lowest number received in year was 1,042 in 1922 and the highest 1,951 in 1929. Anonymous complaints were the most frequent.[51]

It must be remembered, of course, that while official union policy was to work with the inspectorate in adopting safe methods of work, individual members or non-unionised workers may have been apathetic or hostile. In 1937 the Amalgamated Engineering Union (A.E.U.) *Monthly Journal* tried to impress on its members the beneficial effect of the inspectors who, according to the journal, were often regarded with suspicion and had much kept from them.[52] Assessing individual workers' attitudes is inevitably anecdotal.[53] There are numerous references in the Annual Reports of the Chief Inspector of Factories to the indifference of workers to safe methods of working. The Chief Inspector commented in his report for 1931 that for 'any substantial reduction in accidents to occur means were needed to deal with the large volume of accidents wholly unconnected with machinery', and he thought the answer was to be found in the Safety First movement.[54] In his report for 1937 the Chief Inspector again recounted instances of workers' 'carelessness or downright negligence' causing accidents: a foreman started up a machine without warning, even though he knew that men were working on it and one man was seriously injured; a man repairing a machine gave a signal for it to be restarted while he was still in a dangerous position and he was caught and killed, and a weaver was killed when a man repairing a loom set it in motion without first checking that the weaver was clear.[55] The findings of the Industrial Health Research Board also emphasised the individual factor in accident causation. On the shop floor there was a long tradition of indifference or resignation to accidents and unhealthy conditions going back to before industrialisation.[56] Most histories of individual unions make little or no reference to safety, health and welfare; this is probably a true reflection of the unions' concerns at this time although unions' official policy probably displayed a greater concern for good working conditions than did the individual members. While accepting the point that one cannot place members of the trade union hierarchy in discrete categories,[57] nevertheless, it would seem that the further up the trade union hierarchy, the greater the interest displayed in safety, health and welfare. Thus the T.U.C. did co-operate with the inspectorate in its educative work and pressure was put on governments to improve legislation and increase the number of inspectors. Perhaps the T.U.C. saw it as part of its role to press for greater state involvement to protect its members, coupled with the knowledge that it did stand some chance of influencing the content of legislation, in a minor way, in the long term. Individual unions tended to take up the question in response to a specific situation which was perceived as directly affecting them, such as a particularly nasty accident. It was more difficult for external pressures, such as the factory inspectorate, to get workers to take up issues which the inspectorate thought to be important to them.

The great disparity between the aims and achievements of inspectors, in part reflected fiscal restraints, but it was also due to the absence of agreement over safety, health and welfare between factory inspectors on the one hand and the Courts, employers and workers on the other.

IV

A number of factors worked to reduce the effectiveness of the inspectorate's work. First, the educative work of the inspectorate was hampered by the environment in which it took place: a factory is geared towards production, not education. (The Industrial Museum provided a more appropriate environment.) Second, the inspectorate was attempting to educate an adult working population and there were difficulties in 'pitching' propaganda at the right level so that it was digestible without being patronising, and organised in such a fashion that it did not place additional burdens on the government's finances.

Third, the enforcement of certain practices by the use of criminal sanctions raised severe problems. Actually detecting breaches of the law could be difficult, especially when visits were often cursory (on average an inspector made four visits every day). We know that there was an occupational bias in the composition of the Bench and we also know that there was an element of bias in sentencing for Factory Act crimes. We do not know how widespread was the bias or whether it was typical of sentencing for other crimes. Further research is required in order to assess how far changes in the Bench had lagged behind social and political change.

Fourth, the industrial circumstances of the 1930s aggravated the difficulty of reducing accidents: an increase in accidents was likely to occur when extensive alterations in machinery and plant were made, necessitated by a changeover to new processes and manufactures, while attempting to maintain production during the transitory period. Often when occupiers installed new plant they waited until the factory inspector visited the factory before providing fencing. In 1936 the Chief Inspector felt that the accident rate increased disproportionately to the number of workers in times of prosperity because there were more inexperienced young workers and workers previously unemployed and long out of practice absorbed into industry, while new machinery with unknown risks had been installed. Conversely, in times of depression working hours were shorter, production slower and possibly the more accident prone section of the population were discharged, leaving a more stable residue.

Finally, it is important not to gloss over the negative effect of the inspectorate's concentration on education. The first medical inspector, Sir Thomas Legge, admitted, 'Until the employer has done everything, and everything is a good deal, to protect the worker, the latter can do nothing to protect himself.' Yet, by emphasising the role of individual workers in causing accidents, the problem of unsafe and unhealthy working conditions was individualised; it implied that the worker was independent and in control of his own safety and health at the workplace, thereby deflecting attention both from those aspects of working conditions, such as industrial diseases and general environmental factors, which were beyond the individual worker's control, and from a structural analysis and hence a more radical solution. In this the inspectorate reflected, rather than challenged, the dominant ideology, which held that well-being derived from individual effort.

In conclusion, our examination of the factory inspectorate shows that caution is needed in equating employers' power with the extent to which employers' organisations were incorporated into the policy making process. Until recently, historians have tended to ignore employers' organisations, but it is important that we do not swing the other way and overemphasise the relative strength of employers' organisations when discussing power within the state. During the 1930s the power of organised employers was apparently curtailed in the policy making process for occupational health and safety,[58] yet employers' influence at the implementation stage remained strong and for this reason, concessions to, and gains won by, the labour movement were often Pyrrhic victories.

Notes

1   Eugene Bardach, *The Implementation Game: What happens after a Bill becomes Law* (Cambridge, Mass., 1977), p. 33.

2   For a discussion of the understaffing and increased workload of the inspectorate see Helen Jones, 'The Home Office and Working Conditions 1914-1940', Ph.D. London, 1983, chapter 6.

3   P.W.J. Bartrip and P.T. Fenn, 'The Conventionalisation of Factory Crime — A Reassessment', *International Journal of the Sociology of Law*, vol. 8 (1980), pp. 175-186, 182.

4   W.G. Carson, 'White Collar Crime and the Enforcement of Factory Legislation', *British Journal of Criminology*, vol. 10(1970), pp. 383-398. See also W.G. Carson, 'The Conventionalisation of Early Factory Crime', *International Journal of the Sociology of Law*, vol. 7, no. 1 (1979), pp. 37-60.

5   P.W.J. Bartrip and P.T. Fenn, 'The Conventionalisation of Factory Crime — A Reassessment', *International Journal of the Sociology of Law*, vol. 8 (1980); P.W.J. Bartrip and P.T. Fenn, 'The Administration of Safety: The Enforcement Policy of the Early Factory Inspectorate', *Public Administration*, vol. 58 (1980), pp. 87-102, 99.

6   W.G. Carson, 'Some Sociological Aspects of Strict Liability and the Enforcement of Factory Legislation', *Modern Law Review*, vol. 33, no. 4 (1970), pp. 396-412, 411.

7   John Baldwin and A. Keith Bottomley (eds.), *Criminal Justice: Selected Readings* (Oxford, 1978), p. 4.

8   A. Keith Bottomley, *Criminology in Focus: Past Trends and Future Prospects* (Oxford, 1979), p. 97.

9   W.G. Carson and Paul Wiles (eds.), *The Sociology of Crime and Delinquency in Britain vol.1. The British Tradition* (Oxford, 1971), p. 190.

10   W.G. Carson, 'White Collar Crime and the Enforcement of Factory Legislation', *British Journal of Criminology*, vol. 10 (1970), pp. 383-398, 392.

11   A. Keith Bottomley, *Criminology* (1979), p. 103.

12   W.G. Carson, *The Other Price of Britain's Oil* (Oxford, 1982), p. 7.

13   Edwin Sutherland, *White Collar Crime* (New York, 1949), p. 47.

14   R.I. Mawby, 'Policing by the Post Office: A Study of Television Licence Evasion', *British Journal of Criminology*, vol. 19, no.3 (July 1979), pp. 242-253.

15   R. Quinney, *The Social Reality of Crime* (Boston, 1970), p. 16 quoted in A. Keith Bottomley, *Criminology* (1979), pp. 11-12.

16   This applies to all aspects of white collar crime. Gilbert Geis and Robert F. Meier, *White Collar Crime* (New York, 1977), p. 17.

17   A. Keith Bottomley, *Criminology* (1979), p. 37.

18   Ibid., p. 38.

19   Eugene Bardach, *Implementation* (1977), p. 66; D.S. Van Meter and C.E. Van Horn, 'The Policy Implementation Process: A Conceptual Framework', *Administration and Society*, vol. 6, no. 4 (1975), pp. 445-488.

20   The inspectorate was concerned not only with accident prevention but also occupational health in the broader sense of cleanliness, sanitation, ventilation and dangerous processes.

21   H. Ball and M. Friedman, 'The Use of Criminal Sanctions in the Enforcement of Economic Legislation: A Sociological View', in Gilbert Geis and Robert F. Meier, *White Collar Crime* (1977), pp. 318-336, 333.

22   This is one of the problems of defining violation of the Factory Acts as 'white collar crime'.

23   P.P.1919 vol.XXII Annual Report of the Chief Inspector of Factories (hereafter Annual Report) for 1918, p. 15.

24   P.R.O. LAB 15/43. Slocock 21st September 1934 to Chief Inspector. Factory Act cases were heard in a Court of Summary Jurisdiction, presided over by a minimum of two magistrates, or one if he/she was a Stipendiary one.

25   P.R.O. LAB 14/234. Cases of dismissal of employees for giving evidence to Factory Inspectors.

26   *Annual Survey of English Law* (1928), p. 211; (1932), p. 251; (1933), p. 244.

27   *The Magistrate* (August 1929).

28   *The Magistrate* (November-December) 1937.

29   H.C. Deb. vol. 268 (6th July 1932), col.468 R. Davies. See also *The Magistrate* (July-August) 1936.

30   P.P. 1947-48 vol. XII Report of the Royal Commission on Justices of the Peace, 1946-48.

31   P.W.J. Bartrip and P.T. Fenn, 'The Conventionalisation of Factory Crime — A Reassessment', loc. cit. p. 181.

32   Helen Jones The Home Office chapters 2, 3, 4.

33   The class bias of the Bench still exists at the time of writing, and there is some evidence that a relationship exists between the composition of the Bench and its sentencing practices. John Baldwin, 'The Social Composition of the Magistracy', *British Journal of Criminology*, vol. 16, no.2 (April 1976), pp.171-174,174.

34   P.P. 1929-1930 vol.IX Annual Report for 1928 pp. 70-71.

35   P.P. 1930-31 vol. XIII Annual Report for 1930 p. 82.

36   P.P. 1929-30 vol.XIII Annual Report for 1928 pp. 73, 82.

37   P.P. 1930-31 vol.XIII Annual Report for 1930 p. 92.

38   P.P. 1929-30 vol.XIII Annual Report for 1929 p. 26.

39   Home Office, *Descriptive Account and Catalogue of the Home Office Industrial Museum and Exhibits*. (H.M.S.O., 1928); Home Office, *Outline Guide to the Home Office Industrial Museum and Exhibits* (1938).

40   P.P. 1931-32 vol.IX Annual Report for 1931 p. 6.

41   P.P. 1931-32 vol.IX Annual Report for 1931 p. 22.

42   The Cotton Spinners and Manufacturers Association. *Agreements reached at conferences between Employers, Operatives and Inspectors, held 1926-28 concerning fencing of machinery, prevention of accidents, and First Aid in cotton weaving factories.* 8th February 1929.

43   For a full discussion see Helen Jones, 'Employers' Welfare Schemes and Industrial Relations in Inter-War Britain', *Business History*, vol. 25 (March 1983), pp. 61-75.

44   P.P. 1932-33 vol.XII Annual Report for 1932 p. 55.

45   P.E.P., *Report on the British Health Services: A Survey of the existing Health Services in Great Britain with proposals for future developments* (December 1937), p. 82.

46   D. Stewart, 'Industrial Medical Services in Great Britain: A Critical Survey', *British Medical Journal*, vol. 2 (29th November 1941), pp. 762-765.

47   T.U.C. Archives T 248 Factory Inspection Committee.

48   Ibid.

49   T.U.C. Archives T 507 146-T 1927-45.

50   United Ladies' Tailors' Trade Union, *Report* (1929).

51   There were four categories of complaint: anonymous, official, operatives and 'other'. Over half of the complaints were

verified. Annual Report 1919-1931 passim (excludes Report for 1921 when reorganisation of the Districts meant certain statistics were not collated).

52   A.E.U., *Monthly Journal* (September 1937).

53   For instance, Melvyn Bragg, *Speak for England* (London, 1978), p. 181; Peter Donnelly, *The Yellow Rock* (London, 1950), p. 128; Maurice Levinson, *The Trouble With Yesterday* (London, 1946), pp. 42-43.

54   P.P. 1931-32 vol. IX Annual Report for 1931.

55   P.P. 1937-38 vol. X Annual Report for 1937 pp. 25-26.

56   John Rule, *The Experience of Labour in Eighteenth Century England* (London, 1981), pp. 74-94.

57   Jonathan Zeitlin, 'Trade Unions and Job Control — A Critique of Rank and Filism', *Society for the Study of Labour History Bulletin*, no. 46 (Spring 1983), pp. 6-7.

58   Helen Jones, thesis chapters 2, 3 and 4.

13 'THE GOLDEN FACTORY'. [1] INDUSTRIAL HEALTH AND SCIENTIFIC MANAGEMENT IN AN ITALIAN LIGHT ENGINEERING FIRM. THE MAGNETI MARELLI IN THE FASCIST PERIOD

Perry Willson

Milan was an important city in the development of industrial health care since it was there that Luigi Devoto opened the world's first occupational health clinic in 1910. The Clinica del lavoro, which inspired the creation of many others all over the world, was founded 'to scientifically study the causes of occupational diseases and improve doctors' knowledge of them: to provide in-patient diagnosis and treatment for workers with either suspected or confirmed cases of industrial diseases: to make periodic inspections of the health of industrial workers in general and especially of those who did particularly unhealthy jobs'.[2] Devoto's work meant that, in some respects, Italy was a pioneer in occupational medicine. In the field of preventive legislation, however, she lagged behind, and insurance against industrial accidents did not become compulsory until 20 January 1904 and even once it came into force this law was ignored by the majority of firms. In 1906, a circular from the Ministry of Agriculture, Industry and Trade noted that the law was 'generally respected very little. When factory inspectors complain the majority of industrialists and entrepreneurs reply that they have never even heard of the legislation.'[3] Meanwhile, in larger firms there was a growing interest in accident prevention and, in 1894, the *Associazione degli industriali d'Italia per prevenire gli infortuni* was formed. This gathered statistics on accidents, inspected factories and distributed propaganda on accident prevention to industrialists, workers and engineering students. At first, it was very active in its attempts to encourage accident prevention, although its missionary zeal seems to have declined with time.[4]

The legislation and the employers' initiative were, to a certain extent, inspired by fear of the strength of the workers' movement around the turn of the century. With the seizure of power by the Fascist Party in 1922, the working class movement was crushed and in the next few years free trade unions were replaced by Fascist *sindacati* and the right to strike abolished. The interest in health and safety did not,

however, totally disappear and in 1934 new legislation came into force
which extended compulsory insurance to cover six industrial diseases:
lead, mercury, phosphorous, carbon disulphide and benzol poisoning
and anchylosis. 1933 saw the creation of INFAIL *(Istituto nazionale
fascista per l'assicurazione contro gli infortuni sul lavoro)* a state organisation
which centralised all the insurance bodies.[5]

The rise to power of Mussolini's party did, however, create problems
for some members of the medical profession, since the destruction of
the free trade unions undermined the relationship of co-operation
which had begun to evolve between certain factory doctors and
workers in the pre-fascist period. For Devoto, this was also a difficult
time. In some respects, the *Clinica del lavoro* was a democratic
institution as it 'represented the acknowledgement of a very important
problem, that is the existence of the unequal position in society of one
class, the working class'. [6] Devoto himself referred to the clinic as 'a
child of the people' [7] and adopted the motto 'equality for all in health,
everyone should benefit from the advances of medical science'. [8] It is
perhaps not surprising, therefore, that 'With the advent of Fascism ...
the Clinic ... fell under a shadow for a while ... it was treated with
suspicion and regarded as a source of demagogy, and the numerous
declarations of affection from the working classes seen as
compromising. As a result, the Clinic was thrown into almost total
isolation'.[9]

In spite of this, the inter-war period was not entirely negative in the
field of industrial health since, as Luisa Dodi Osnaghi has argued, [10]
the medical profession was becoming increasingly interested in this
question and particularly in the problem of toxicity, as the chemical
industry was expanding and new processes and materials were being
introduced. Certain medical experts were also beginning to look
beyond the narrow definition of industrial health which simply saw it
in terms of dangers in the workplace. Factory doctors, they stressed,
should take a broader view of workers' health and consider both the
home and work environments.

Despite these advances in medical awareness of occupational health
questions, conditions in many industries were very bad. This was to a
certain extent due to the particular way in which American
management ideas were introduced into Italian industry. Henry
Ford's mass production of cars and Frederick Taylor's experiments
with 'scientific management' aroused a great deal of interest in Italy.
Numerous articles appeared in management publications discussing
the applicability of these ideas but to implement them necessitated a

complete reorganisation of the factory which most firms were not prepared to undertake. Instead, they simply copied a few selected aspects of the American form of management, such as the use of piecework and a faster rate of production. Technological innovation remained limited and wages were low.[11]

This implied a whole new range of industrial health problems, but instead of looking for new solutions the contemporary analysis tended to blame the victims and in the opinion of many occupational health experts, 'with increasing frequency workers were inflicting injuries on themselves in order to obtain whatever small compensation the law provided ... this accusation that injuries were self-inflicted was the subject of a propaganda campaign during the whole [interwar] period, which became even more insistent during the debate about the reform of the 1904 legislation'.[12] The employers' organisation, the *Confindustria,* for example, claimed, in 1930, that only between 8 and 30 per cent of all accidents were caused by machinery, whereas 70 to 75 per cent could be attributed to the workers themselves. [13] The logical conclusion of this was that if workers caused accidents then they should be sacked. Giulio Sapelli argues that many employers did not bother to introduce safety measures since no unions safeguarded the workers' interests and he sums up the policy of the majority of firms as the, 'Selection and elimination of unhealthy workers and their replacement with younger elements whose health had not yet been impaired by factory work'. [14]

In one sector, at least, this seems to be an accurate description of the situation. As a recent study by Bruna Bianchi [15] shows, in the textile industry health and safety standards were appalling. This was still a major sector in the Italian economy. In rayon production, for example, Italy was second only to the U.S.A. and she was the world's largest exporter. Capital investment in new technology was minimal and productivity was raised mainly by increasing the pace of work to an intolerable level. Protective equipment, such as ventilators, was often lacking and the workforce, which was largely female, had a high turn-over rate because worn-out workers were often sacked and replaced. Conditions in the rayon industry were especially bad and many workers were exposed to toxic carbon disulphide fumes. In these cases, however, those who were sacked quickly were often the lucky ones since, in the advanced stages of this poisoning, the victims behaved as if they were mad. Many were shut up in mental hospitals despite the fact that medical journals had published research proving the link between exposure to these fumes and symptoms of madness.

This method of shutting away the evidence was further facilitated by the fact that many of the victims were women. In the Snia Viscosa factory in Venaria Reale, for example, doctors refused to take the symptoms of young factory girls seriously since they considered that women had 'naturally' hysterical temperaments. In another factory the symptoms of an entire shop were dismissed as collective hysteria. Thus, all the rayon workers were vulnerable since no union or right to work protected their interests but the women among them were the most defenceless due to the myth that women were essentially irrational creatures prone to hysteria.

Unfortunately, there is only a very limited amount of comparable research into industrial health in other sectors of Italian industry in the Fascist period. Although a number of important studies of Italian firms have been published, they have tended to ignore the question of industrial health, to a great extent due to the fact that adequate source materials are very scarce on this subject. There is still a need, therefore, for more research into the situation in specific factories. The case of the Magneti Marelli is of special interest, although far from typical.

The Fabbrica Italiana Magneti Marelli was founded in Milan in 1919 with capital jointly provided by the Turin car manufacturer Fiat and a heavy engineering firm the Ercole Marelli. Initially set up in response to the new market for magnetos created by the expansion of aviation during the First World War, the firm quickly extended its range of products to include numerous electrical parts for cars, aeroplanes and trains as well as radios and military equipment. The rising demand for radios, the links with Fiat and a number of important state orders meant that this firm was in the fairly unusual situation of manufacturing for a steady and expanding market.[16] The autarky policy (which aimed to make Italy's economy self-sufficient) was also important in its growth since this gave the Magneti Marelli a monopoly on the Italian market for many of its products. Certain Italian firms, such as car manufacturers, were specifically directed by the Fascist government to use Magneti Marelli parts rather than imports. The original small plant which opened in Sesto San Giovanni, an industrial suburb of Milan, was added to and by the Second World War there were four plants in Sesto and Milan alone and a number of others scattered in various parts of Italy. This rapid growth should not, however, be ascribed to secure markets alone since the Magneti Marelli made a serious attempt to introduce Fordist and Taylorist ideas. The first conveyor belt assembly line, for example, was already functioning by 1924. The only other firm with a comparable approach to the organisation of production in this period was Fiat. The Magneti Marelli management, furthermore, took a special interest in 'scientific personnel management'. This innovative and pioneering approach was also applied to the health and safety policy.

The variety and complexity of production of this firm created a huge range of potential health hazards. For example, products such as radios and car parts required dangerous cutting machinery, batteries involved toxic chemicals and the foundry plant presented numerous risks with its furnaces and sandblasting equipment. Despite this, the evidence that has survived[17] of accidents and health problems is largely restricted to workers in the cutting shop, who frequently injured their hands, and those who contracted silicosis from handling chemicals or working on the sandblasting equipment. The fact that accidents do not seem to have occurred so frequently in other jobs suggests that the health and safety policy may have had some effect.

In June 1927, the Magneti Marelli became the first factory in Italy to set up a safety committee. Sapelli[18] argues that, whereas these committees fought the industrial accident rate seriously in their country of origin, the USA, in Italy their role was very limited, concentrating on propaganda to educate workers to be more careful, and watching out for those who were unsuitable for their jobs in order to dismiss them. This may have been true in other industries but it would be an over-simplistic assessment of the Magneti Marelli Safety Committee.

This consisted of a group of foremen, workers and technical and managerial staff.[19] The factory doctor, the social worker and other representatives of the firm's extensive welfare services were members, although, by the 1930s, their role seems to have diminished to an advisory one. The Committee worked in co-operation with all sections of the factory including the Organisational Department (responsable for the 'scientific organisation of production') . The exact composition and structure of this Committee was modified on a number of occasions but two important features remained constant. One was that, although this was ostensibly a mixed committee of workers and managers, the workers remained essentially subordinate to the management. The other point is that, in a factory where over a third of the workforce was female, the Safety Committee was composed entirely of men with the sole exception of the factory social worker. There is no evidence that a single female member of the workforce participated, in spite of the fact that many accidents occurred in female-designated jobs, such as cutting. This meant that, despite appearances, the Safety Committee was not a power-sharing exercise. Yet workers were not included for simple window-dressing, they had a role to play. To understand this, we need to look at how the Safety Committee worked.

When accidents occurred, the Committee investigated. Sometimes this simply meant collecting a report from the foreman but frequently it entailed carrying out a detailed investigation or even getting the help of the worker to reconstruct the event. In this way, the Committee gathered statistics on the seriousness and frequency of accidents as well as investigating their causes. Recommendations were usually made on the prevention of future accidents, such as the introduction of protective clothing or shields. In drawing up the report the Committee worked in co-operation with various technical and managerial departments of the factory.

For example, when Adelina Sigurta, a worker in the magnet making shop, cut her finger on a grinder, the Safety Committee made a thorough investigation. Not only was this particular accident taken into consideration but also the records of previous similar cases were examined in the search for a solution. The report concluded that another similar accident could not be prevented in the future by simply installing better protective equipment. The matter was, therefore, referred to the Organisational Department which proposed three solutions. One of these suggestions has, unfortunately, been lost and the second was simply a mechanical adaption but, interestingly, the third proposed an actual alteration of the cycle of the work-process, so that the order in which certain tasks were carried out was modified.[20] This constitutes a broad and innovative approach to accident prevention.

The Committee was also involved in an active propaganda campaign. The workers who were members were important in this since they were drawn from every shop of the factory and could therefore relay new ideas on health and safety to their work companions. A special noticeboard was got up with photographs, stories and advice, featuring slogans such as, 'A pay packet is always bigger than injury benefit'[21] or 'A word of advice to a workmate costs nothing and can save him from an accident'.[22] This also presented statistics and information about accidents that had already happened and accounts of the Safety Committee's activities. To encourage interest, humorous stories were invented about a fictitious character called '*Infortunello*' who had every conceivable form of accident and competitions were run to find the work team with the lowest accident rate. Safety booklets were produced for the workforce and visits were organised to exhibitions. Numerous articles were published in the factory newspaper.

One of the aims of this propaganda was to stress how much the firm was doing for the workforce. The articles in the factory newspaper, for example, were generally focused on the *active* role of the firm in fighting the accident rate and industrial disease. Less attention was paid to the dangers presented by particular types of machinery, except in the context of describing protective innovations. The management was interested, therefore, not only in protecting the workers' health, but also in convincing them that their welfare was cared for. The campaign also tried to encourage the workers to participate personally in the fight against industrial accidents by taking greater care during their work, suggesting new ideas for accident prevention or serving on the Safety Committee. Only a tiny fraction of the workforce was illiterate [23] so that the majority of them had access to this propaganda, but it is difficult to assess whether it had any actual effect, although there is some evidence that workers did make suggestions about making machinery more secure. [24]

The Magneti Marelli Safety Commitee saw industrial health as a complex issue, involving a mixture of technical, organisational, environmental and human factors. The last two were also the concern of the Industrial Psychology Department, which was another American-inspired innovation . This Department had been created to apply 'scientific' techniques to the firm's personnel management.

In interwar Italy, there was a certain amount of interest in industrial psychology, especially among academics such as Agostino Gemelli (Catholic University of Milan) and Mario Ponzo (University of Turin). Numerous articles were published in management journals and in November 1932 a free industrial psychology advisory service for industrialists was set up in Rome by ENIOS *(Ente nazionale italiano per l'organizzazione scientifico del lavoro)*, an organisation which had been created by the government to encourage the introduction of scientific management methods in Italy. However, these efforts were largely fruitless and only a handful of firms even attempted to apply these ideas. [25] With regard to the vast majority of Italian firms, Sapelli was probably correct when he asserted that, 'In spite of the remarkable efforts of men like Agostino Gemelli, there is no evidence of an effective introduction of industrial psychology'.[26] The Magneti Marelli, however, actually set up a special department to introduce these new ideas and, as Mario Fossati noted in 1927, in this factory new workers were, 'for the first time in Italy, scientifically recruited, which means that a worker is only taken on if he fulfils certain requirements and he is assigned to the shop to which he is most suited both mentally and physically'. [27]

Run by the factory doctor, Annibale Correggiari, the Industrial Psychology Department contributed to the health and safety campaign in two ways. Firstly, it developed a 'rational' method of selecting new employees and, secondly, it studied the work environment to assess which factors might give rise to health problems in the long run. This 'rational' selection procedure consisted in a series of tests which each applicant for work had to undergo. Specialised equipment was used, such as the Beyne and Bettagne apparatus which measured the speed of reaction to a visual stimulus. The worker tested had to interrupt an electric current as soon as a light became visible. A mechanical pen automatically recorded the speed. Other equipment measured factors such as the level of tiredness experienced by a worker at different stages of the day, the sensitivity to touch of the fingers, physical strength or accuracy in drawing a line exactly half way between two others. [28]

Workers were tested in this way principally to raise output by improving the 'quality' of the workforce but this was also seen as a way of preventing accidents since workers were only employed if they could perform a task fast and efficiently enough. The ideal score for each of these tests was determined by testing the existing workforce. The recruitment strategy may have cut the accident rate but it also aroused the suspicions of the workers, who feared that it could be used to eliminate the weak, ill or ageing who often desperately needed work. Umberto Quintavalle, one of the directors, specifically tried to refute these allegations in an article in 1928, where he argued that although the psychological test had been criticised as a method of eliminating workers who became ill, it was not actually used in this way but instead ensured that workers were not given jobs too taxing for their capacities. [29] Evidence from the factory archive does, however, reveal that the workers' fears were not unfounded. In 1933, a memorandum to the *Amministratore Delegato* (managing director) concerning the sacking of a worker notes that, 'At the moment the management is reviewing the workforce to eliminate those employees who are the least productive because of their age or temperament, and replacing those dismissed with the young and energetic'. [30] This particular worker was sacked simply because he was too old.

This testing procedure was aimed at fitting the worker to the needs of machinery and the organisation of the work-process, in contrast to some of the work of the Safety Committee which modified machinery and the work environment to fit the needs of the workforce. The Industrial Psychology Department was also, however, concerned with the environmental causes of accidents and industrial diseases. Light, heat and ventilation were all monitored and it was attempted to improve the position in which each worker had to sit or stand. For the measurement of light, for example, a special machine, the 'Luxmetro', was used. A table was drawn up of the number of 'Lux' units necessary for each task and the machine was carried around the factory making checks.[31]

The work of the Safety Committee and the Industrial Psychology Department demonstrate that this firm took the question of industrial health very seriously. It is not easy, however, to evaluate to what extent this campaign was actually successful. The only statistics that we have been able to find show the effects of the accident prevention policy in the first two years after the foundation of the Safety Committee.

According to these figures, published in an article by Correggiari in 1930,[32] the accident rate fell substantially in 1927 and 1928 in comparison with 1926 (Table 1). This reduction was not, however, reflected in the amount of production time lost due to the granting of temporary sick leave following accidents.

Table 1: Number of Accidents per 1000 Hours Worked, 1926 to 1928. (1926 = 100)

1926	100
1927	94
1928	76

Source : A.Correggiari, 'La prova dei fatti', *Sprazzi e Bagliori*, gennaio 1930, p. 45.

The number of days wasted in this way remained at roughly the same level during the three years under consideration. Correggiari dismisses these figures as unimportant and maintains that they were most probably simply due to the fact that some of the staff of the medical centre were more generous than others when they granted sick leave.[33] An alternative explanation might be, however, that the seriousness of smaller accidents was increasing, so that longer periods of time were necessary for recovery. On the other hand, the seriousness of accidents which caused permanent disabilities seems to have been diminishing. Judging the seriousness of an accident according to a table drawn up by the American Standardisation Commission, Correggiari demonstrates that the firm's expenditure in permanent disability compensation was greatly reduced. Since the compensation for permanent disability represented a much bigger expense than the loss of working days in temporary sick leave he was able to show that the overall result of the accident prevention policy was an impressive saving for the firm. (Table 2)

Table 2: Cost of Industrial Accidents per 1000 Hours Worked, 1926 to 1928. (1926 = 100)

1926	100
1927	78
1928	70

Source : A.Correggiari, 'La prova dei fatti', *Sprazzi e Bagliori*, gennaio 1930, p. 46.

Although no figures are quoted for 1929, since at the time of writing they had not yet been compiled, Correggiari maintains that the figures for the first ten months promised that the accident rate would prove to be as low or even lower than the 1928 level. [34]

Without further information, it is not possible to evaluate these figures properly, but they do seem to suggest that the accident prevention policy did achieve a reduction in both the number and seriousness of accidents in the first years of its existence. Unfortunately, we have no comparable figures for the 1930s, nor any indication of whether industrial diseases became less common.

Whether successful or not, nevertheless, these initiatives add up to a serious health and safety campaign in comparison to many other contemporary firms and they testify to the fact that the Magneti Marelli management and the factory doctor Correggiari were prepared to experiment with American ideas that had not yet been tried in Italy. The factory archive does, however, reveal one striking exception to this generally harmonious picture. This firm was keen to present the image of a caring management doing everything possible to safeguard the workforce's health, but the question of silicosis demonstrates that they were not willing to accept any questioning of their ability to do so. The cases of silicosis also suggest that the health and safety policy may have altered with time and the early seriousness and enthusiasm may have given way to a more strong arm policy. It is, however, very difficult to be certain about this since the absence of documents relating to the earlier period may simply mean that they have been lost.[35]

In Italy, silicosis had traditionally been seen as a miners' disease and it was only slowly recognised that it could also be an occupational health risk in certain engineering jobs. Since the Italian mining sector was small this was considered a rare disease and the 1934 legislation which made compensation automatic for six industrial diseases omitted silicosis from the list.[36] In spite of this, a number of Magneti Marelli workers did attempt to get compensation in the late 1930s and early 1940s.

Most of these sued after the firm sacked them for reasons of ill-health.[37] Only one record survives of a worker being paid compensation and the sum was fairly small — equivalent to about eighty days' pay.[38] The firm won certain cases by arguing that the problems that these workers had with their lungs were due to other causes, such as bronchitis whose origin was outside the workplace. [39] Yet medical journals in the 1930s show that that the link between silicosis and sandblasting or any other job carried out near it, was understood. Many contemporary medical experts were aware of these dangers and, furthermore, they considered that the use of sealed-in workcabins and ventilators, such as those cited by the Magneti Marelli in its defence, was insufficient protection. The firm was willing to pay high fees for lawyers and professional experts which suggests that it was less concerned about the immediate expense of paying compensation to the particular worker concerned, than about the danger of setting a precedent. This may mean that the eight surviving records of cases of workers who took the firm to court were only a fraction of the potential number. The case of Francesco Galbiati [40] demonstrates the atmosphere in which workers tried to get compensation.

Galbiati worked, in 1935, in the Heat Treatment Shop, next to a vat of potassium cyanide. Fumes from this permanently damaged his lungs, as was recognised by the factory medical officer Fulghieri, who recommended that he should be transferred on to other work. But Galbiati wanted compensation since his health was ruined and he took the firm to court. He was unable to get compensation, however, until after the fall of Fascism and the Liberation, when the firm was temporarily run by a managing director appointed by the National Liberation Committee — Brasca. In a letter to Brasca in 1946,[41] Galbiati describes how the Personnel Manager of the firm tried to prevent the court case by promising a friendly settlement. No money ever appeared and, as Galbiati continued to push for justice, the promises of a cash payment turned to threats and he had to drop the matter. By this time, it was 1942 and workers who caused trouble could be mobilised into the army.

These cases of silicosis underline the essentially authoritarian nature of the firm's policy. The management took the issue of health and safety very seriously but they would not allow their competence in this matter to be questioned by the workforce. This was not difficult in a regime that had destroyed trade unions and favoured employers rather than workers, a situation that was only further exacerbated by the war. Further cracks appear in this apparently benevolent policy, if we examine the management's motivation for taking such as interest in industrial health. To understand this, we need to look first at the relationship between industrial health and scientific management.

Henry Ford was interested in industrial health and safety, but not simply for charitable reasons. He maintained that not only was it socially desirable to protect the health of the workforce but also that it could help keep productivity high. 'One point [he wrote] that is absolutely essential to high capacity as well as to humane production, is a clean, well-lighted and well-ventilated factory'. [42] In his view, the social benefits of an industrial health and safety policy were an integral part of scientific management. In spite of this, his policy retained paternalistic and authoritarian characteristics since it saw workers essentially as children who were unable to look after themselves.

The Magneti Marelli, strongly influenced by American ideas, saw health and safety as more than simply a question of protecting the workers' well-being. Correggiari stressed this in 1929, when he claimed that, in this firm, 'accident prevention is based on the principle of *rational organisation*. In the eyes of both the management and the workers the link between scientific management and accidents is perfectly clear.' [his italics] [43] He does not, however, clarify exactly what he means by this and we need to look more closely at the nature of this link.

To a certain extent it is true that a dynamic health and safety policy was actually necessitated by the American-style organisation of production which created new and unprecedented hazards. Production tempos were much faster and at this accelerated pace workers could be under constant stress and therefore at risk. Umberto Quintavalle recognised this when he wrote, in 1928, that, 'the faster work tempo imposed by modern rationally organised production only worsens the [health and safety] situation, since the nervous tension and increased physical effort required undermines the worker's resistance unless we also introduce appropriate preventative measures '. [44] Not only the pace but also the type of work could cause problems since, to some extent, scientific management meant the division of work into repetitive and essentially boring tasks, which dulled the workers' senses and made them less careful. There were further dangers inherent in the rapid technological innovation; constantly changing machinery and new processes meant unknown health risks. In contrast to some other firms who tried to deny the existence of risks specifically created by the new approach to management, all of these problems were taken seriously by the Magneti Marelli.

Scientific management not only caused health risks but also, conversely, facilitated an innovative approach to the problem. The tools of scientific management could, themselves, be used in the fight against industrial disease and accidents. Thus, it could be claimed that the health policy was, in itself, 'rationalised'. New dangers were tackled with innovative weapons, which meant not only the American-inspired Safety Committee, but also the Industrial Psychology Department with its 'rational' recruitment policy. One reason why this aspect was strongly emphasised by the management was the fact that the rapid changes inherent in scientific management had to be imposed upon the workforce. In this situation, it was important to attempt to present technological and managerial innovations as beneficial to all.

Furthermore, scientific management enabled the firm to see the industrial health policy not as a cost and burden but rather as a self-financing measure. At first, this was not mentioned and an article in the factory newspaper, [45] in 1925, claimed that workers health was simply a question of the benevolence of the employers. However, this attitude was soon abandoned and various articles on the health and safety policy emphasised that it paid for itself, since accidents which disrupted the flow of work were costly, sick leave and compensation meant lost production time and expense and healthy workers were more productive. In fact, as Correggiari noted, in 1930, the management were not willing to spend any more than they calculated would be returned in this way. 'It may seem [he wrote] that these provisions are too expensive and sometimes have an unfavourable effect on productivity: as far as cost is concerned, it should, of course, always be kept in proportion to the increased profit yielded, otherwise it would be uneconomic.' [46] He does, however, say that, in his opinion, there was not often a conflict between the needs of production and the needs of health, so that this strict budgeting rarely obliged the firm to scrimp on protective measures. Furthermore, in the same article, he goes on to argue that greater safety protection could actually raise productivity and overall profits since, 'dependable accident prevention measures eventually have a psychologically favourable effect on the worker: relieved of the danger which previously threatened him he can dedicate himself more calmly to his task, feeling himself in safety he speeds up his pace of work and sometimes *solely due to the safety measures* there is a net increase in production' [his italics].[47]

The fact that this policy effectively cost the firm nothing throws a new light on the health and safety campaign and explains, to a great extent, why shrewd businessmen like the Quintavalles devoted such time and resources to a policy which at first sight appears to have been all to the benefit of the workforce. There were, however, further gains to be reaped from this type of policy.

We can quickly dismiss as over-simplistic, the claim that the concern for health and safety was largely due to the true Fascist nature of the management. One article in the factory newspaper even maintained that the interest in preventing industrial accidents was principally motivated by the Fascist demographic campaign, since injured workers could not fight as soldiers, 'Amongst the permanently injured the level of handicap varies, but the statistics suggest that at least a third of these would no longer be capable of serving actively as fighting troops in defence of the Motherland; when added to the number of deaths this amounts to an annual loss of about 18,000 men, the size of an entire Division, which is far from negligible ... ' . [48] Although we have no cause to question the management's alliance to the regime, it seems highly unlikely that this approval of the Fascist demographic policy (which aimed to increase the birth-rate for nationalistic reasons), was a serious factor in the decision to create an active health and safety campaign. This kind of claim was probably simply an attempt to make political capital out of a pre-existing policy.

Another aspect of Fascist policy is, however, much more likely to have been taken into serious consideration. The Fascist Labour Charter *(Carta del Lavoro)* had outlawed strikes, promising that the regime would help workers and employers to co-operate productively and peacefully to build a happy and prosperous future. This was the much vaunted policy of 'class co-operation ' which was supposed to replace the old destructiveness of class conflict. In many firms this tended to mean that employers were free to impose heavy wage cuts, huge increases in the pace of work and redundancies. The newly formed Fascist *sindacati* (which replaced the trade unions) could do no more than slightly cushion the effect of this onslaught on the standard of living of the working class. Strikes did occur at times and although they could not succeed they caused disruption. One example of this was the series of strikes among female textile workers in Legnano near Milan, in 1931, where battles occurred with the police.

The problem of making 'class co-operation' work was especially relevant to this firm since three of its plants were situated in Sesto San Giovanni, which grouped most of Milan's large industry. A working class stronghold, it was a focus for anti-Fascist activity. In 1944, for example, a secret Communist report commented that Sesto San Giovanni was,

> undeniably the area [in Milan] where the mass support for the anti-Fascist struggle is strongest, not only regarding the influence of our party and its membership level, but also in the number of actions carried out, the funds collected, the aid to the families of political prisoners, etc... This is the key zone from which all of Milan's workers take their cue before beginning any action.[49]

The Fascist blackshirts had destroyed the political organisations of the working class, but more subtle methods were needed to win their actual co-operation and the Magneti Marelli management were not shy to point out that taking care of the workforce's welfare could also be a way of creating a consensus or, at least, of winning their employees' passive acquiescence. Thus, an article, in 1925, noted that the firm was willing to accept the so-called sacrifices a welfare policy entailed since, 'over and above the moral satisfaction we quite rightly feel in caring for those who have helped the firm to its current greatness, we are also compensated by the fact that the workforce feels a greater bond to the firm... '. [50] Furthermore, willing workers were more productive.

> The strength of an industry like ours lies to a great extent in the goodwill of the workers; a willing worker raises productivity which benefits both production and his own interests; this creates a better and more understanding relationship between worker and firm so that they can count on each other in times of need.[51]

It is, of course, extremely difficult to tell whether this attempt to create a spirit of consensus did actually work; whether, in fact, the workforce did feel 'goodwill' towards the firm. Neither is it possible to be certain to what extent the health and safety campaign was successful in fighting the accident rate and industrial disease, apart from the promising results of the first two years. It is clear, however, that a serious attempt was made to achieve both of these aims, in contrast to many of the firms descibed by Bianchi and Sapelli. The Magneti Marelli was, in many ways, unusual, a rare example, in interwar Italy, of a firm trying to implement Fordist and Taylorist ideas and its health and safety campaign, as carried out by the Safety Committee and the Industrial Psychology Department, constituted an interesting and innovative policy.

Notes

1   The Magneti Marelli was described as a 'golden factory' *(una fabbrica d'oro)* by a former Winding Shop worker employed there from 1933. Interview with Maria G. December 1983.

2   L. Devoto, *La Clinica del lavoro di Milano 1910-1929* (Milano, 1929), p. 13.

3   R. Romano, 'Gli industriali e la prevenzione degli infortuni sul lavoro (1894-1914)', in M.L.Betri and A.Gigli Marchetti (eds.), *Salute e classi lavoratrice in Italia dall'Unita al fascismo* (Milano, 1982), p. 145.

4   Ibid., p. 143.

5   For more information on social insurance legislation see, A. Cherubini, *Storia della previdenza sociale in Italia 1860-1890* (Roma, 1977) and G.Bronzini, 'Legislazione sociale ed istituzioni corporative' in *Annali Feltrinelli 1979-80* (Milano, 1981).

6   A.Carbonini, 'Luigi Devoto e la clinica del lavoro di Milano' in

Betri and Gigli Marchetti, *Salute e classi lavoratrice*, p. 515.

7   L.Devoto, 'La clinica delle malattie professionali degli Istituti clinici di Milano', in *Atti del 1 Congresso internazionale per le malattie del lavoro*, cited in Ibid., p. 515.

8   L.Devoto, *I venticinque anni della clinica del lavoro di Milano* (Milano, 1935), p. 19.

9   L.Devoto, 'Una disciplina italiana e i trenta anni del suo giornale', *La Medicina del Lavoro*, (1931), pp. 479-480. Also cited by L. Dodi Osnaghi, 'Aspetti della condizione operaia e della nocivita attraverso le riviste di medicina del lavoro' in *Annali Feltrinelli 1979-80*, p. 232.

10 Dodi Osnaghi, 'Aspetti', p. 234.

11 G.Sapelli, *Organizzazione, lavoro e innovazione industriale nell'Italia tra le due guerre* (Torino, 1978), passim.

12 B.Bottiglieri, 'Razionalizzazione del lavoro e salute operaia tra le due guerre : l'atteggiamento del sindacato e del governo' in Betri and Gigli Marchetti, *Salute e classi lavoratrice*, p. 871.

13 Sapelli, *Organizzazione*, p. 375.

14 Ibid., p. 372.

15 B. Bianchi, 'I tessili: lavoro, salute, conflitto' in *Annali Feltrinelli*.

16 G. Consonni and G.Tonon, 'Milano: classe e metropoli tra due economie di guerra', in Ibid., p. 422.

17 Various letters and memoranda concerning industrial health were found in the Archivio del Personale della Magneti Marelli (hereafter APMM), which contains the record files of all the workers employed since 1919. Information such as grade, date of birth, job etc. are recorded in these files, and some also include letters and other documents. This material, which was kindly made available to me by the Personnel Department of the Magneti Marelli, is being studied as part of a much wider research project, looking at the history of the firm in the interwar period, with special reference to women workers.

18 Sapelli, *Organizzazione*, p. 377.

19 Various articles in the factory newspaper, *Sprazzi e Bagliori* ('Sparks and Glows', hereafter *SB*) describe the Safety Committee. For example, A. Correggiari, 'Esperimenti Antifortunistici', *SB* , gennaio 1929, pp. 25-31.

20 Unsigned memorandum, [1935], *APMM* fasc. Sigurta Adelina.

21 'In casa nostra', *SB* , maggio 1929, p. 51.

22 Ibid.

23 The level of education is shown on most of the APMM files.

24 Both articles in *SB* and notes in APMM mention workers who have suggested safety modifications.

25 Sapelli, *Organizzazione*, pp. 361-2.

26 G. Sapelli, 'Formazione della forza lavoro e psicotecnica nell'Italia fra le due guerre mondiali', *Quaderni di Sociologia*, no. 1(1977), p. 21.

27 M. Fossati, 'La selezione del personale in una azienda industriale di grande importanza '(prima parte), *SB*, giugno 1927, p. 24.

28 Fossati 'La selezione'(seconda parte) *SB*, luglio 1927, pp. 21-22.

29 U.Quintavalle , 'L'Industria e la lotta contro la tubercolosi ', *L'Assistenza sociale nell'industria*, nov.-dic. 1928, p. 8.

30 Letter with illegible signature, 24 March 1933. *APMM*. Fasc. Russo Francesco .

31 'Per difendere gli operai', *SB*, giugno 1929,p. 52.

32 A.Correggiari, 'La prova dei fatti ', *SB*, gennaio 1930, p. 45.

33 Ibid.

34 Ibid., p. 47.

35 The materials contained in the *APMM* are generally much richer for the 1930s than for the preceding period.

36 Dodi Osnaghi, 'Aspetti' ,p. 274n .

37 *APMM* fasc. Imberti Aristide.

38 E.g. the cases of Magni Carmela and Moretti Ambrogio, *APMM* .

39 Dodi Osnaghi, 'Aspetti', p. 277.

40 *APMM,* Fasc. Galbiati Francesco.

41 Letter to P.C. Ing. Brasca, signed Galbiati Francesco, 12 dicembre 1946, *APMM,* fasc. Galbiati Francesco.

42 H. Ford, *My Life and My Work*, new edn (London,1924), p. 113.

43 Correggiari, 'Esperimenti Antifortunistici', p. 25.

44 Quintavalle,'L'Industria e la lotta', p. 8.

45 'Assistenza sociale', *SB*, luglio 1925, p. 11.

46 Correggiari, 'La prova dei fatti', p. 46.

47 Ibid., p. 47.

48 Vico d'Incerti, 'La lotta contro gli infortuni', *SB*, marzo-aprile 1938, p.37.

49 'Rapporto sulla situazione organizzativa della Federazione di Milano, Milano, 20 aprile 1944', in A.Scalpelli, *Scioperi e guerriglia in Val Padana (1943-45)* (Urbino, 1972), p. 95.

50 'Assistenza sociale', p. 11.

51 Ibid.

INDEX